620. 23 ROBERTS
Noise control in the
built environment.

25-00

D1587815

This book is to be returned on or before
the last date stamped below.   3 1 JAN 1989

RTC

K04080

NOISE CONTROL IN THE BUILT ENVIRONMENT

Noise Control in the Built Environment

Editors

John Roberts
and
Diane Fairhall

Gower Technical

© Gower Publishing Company Limited

All rights reserved. No part of this publication may be reproduced, stored in a retrieval system, or transmitted in any form or by any means, electronic, mechanical, photocopying, recording, or otherwise without the prior permission of Gower Publishing Company Limited.

Published by
Gower Technical,
Gower Publishing Company Limited,
Gower House,
Croft Road,
Aldershot,
Hants GU11 3HR,
England.

Gower Publishing Company,
Old Post Road,
Brookfield,
Vermont 05036,
USA.

British Library Cataloguing in Publication Data

Roberts, John
  Noise control in the built environment.
  1. Noise. Control measures
  I. Title
  620.2′3

693 . 834
RoB

973700

620.23
RoB

2651

ISBN 0 566 09001 5

Printed and bound in Great Britain by
Anchor Press Ltd, Tiptree, Essex

Contents

List of illustrations   vii

Introduction   xi

1   Some theoretical considerations   1

  M. L. Vuillermoz

  Vibrating systems – the wave equation – characteristic impedance
  – sound intensity – sound radiation – influence of the enclosure
  – sound radiation into a room – absorption of sound in a room
  – sound decay and reverberation – sound transmission through
  barriers – working formulae and the decibel notation – further
  reading

2   Units, instrumentation and measurement   23

  K. Scannell and P. Colgrave

  Units and descriptors – acoustic measurements – instrumentation
  – references

3   Environmental noise and vibration   47

  P.T. Freeborn and S.W. Turner

  Noise rating – road traffic noise – aircraft noise – railway noise
  – construction noise – industrial noise – entertainment noise –
  references

4   Sound insulation between dwellings   89

Stephen Rintoul

Transmission paths and mechanics – assessing sound insulation
– insulation by design – constructional techniques – remedial works
– trouble shooting – legal considerations – references

5   Noise control within the industrial environment   137

John Roberts and Bridget Shield

Reasons for noise control – noise control at the design stage –
remedial noise control – reduction of noise at source – reduction
of noise by enclosures, partitions and screens – noise reduction
by acoustic absorption – reduction of noise through good management – conclusions – references

6   Hearing conservation programmes   168

Roger Wills

Criteria – hearing conservation programmes – hearing protection
– references

7   Noise and vibration in building services   192

Diane M. Fairhall

Noise in air distribution systems – noise in water distribution
systems – references

8   Applications for active attenuation   226

H. G. Leventhall and John Roberts

The theory and applications of active attenuation – limitations of
active control systems – loudspeakers for active attenuation – the
prospects for active attenuation – where can active attenuation
be used? – references

Appendix   239
Vibrating systems – the wave equation – the velocity of sound – sound
intensity – sound radiation

Index   247

Illustrations

FIGURES

1.1	Representation of a mass on the end of a spring	2
1.2	Resonance curve showing half power points	3
1.3	Representation of a sound wave	5
1.4	Sound transmission through a barrier	17
1.5	Flexural waves and the coincidence effect	19
2.1	Equal loudness level contours	24
2.2	Sound levels exceeded for the stated percentage of the measurement period (L_n)	28
2.3	Precision sound level meter	30
2.4	Sound intensity analyser	35
2.5	Sectional view of a capacitor microphone cartridge	36
2.6	Schematic accelerometer configurations	37
2.7	Schematic diagram of sound level meter	40
2.8	Graphic level recorder	43
3.1	Average annoyance vs. Noise exposure from aircraft	49
3.2	Variation in noise level with time at two different sites	52
3.3	Frequency spectrum of the traffic noise shown in Figure 3.2	53
3.4	Cumulative distribution of traffic noise shown in Figure 3.2	54
3.5	Dissatisfaction with traffic noise vs. L_{A10} ((18hrs) in dB(A)) averaged hourly between 0600 and 2400 hours	55
3.6	(a) Free field measurement of traffic noise	55
3.6	(b) Facade measurement of traffic noise	55
3.7	(a) All dwellings have a sight of the road; (b) Some dwellings and all of the courtyard are shielded from the road	61
3.8	Transmission paths for traffic induced vibration	63
3.9	Number of complaints with change of L_{Aeq} noise level from pop concerts	82
3.10	(a) Poor internal layout; (b) Good internal layout	83
4.1	Transmission paths	90
4.2	Layout and design zoning	101
4.3	Concrete floors	111
4.4	Platform floor	113
4.5	Raft floors	114
4.6	Secondary independent ceiling	115
4.7	Panel systems for walls	116
4.8	Ducts for service pipes	120
4.9	Wall/floor junctions	122
4.10	Panel and duct junctions	122

5.1	Examples of good and bad layouts for a typical teaching department	139
5.2	High pressure exhaust muffler	143
5.3	Silencer made from pipe fitting and filler	144
5.4	Burner silencer	145
5.5	Acoustic enclosure	147
5.6	Construction of wall tumbler enclosure	151
5.7	Acoustic booth	152
5.8	Cross-section of a typical double-leaf partition	158
5.9	Construction of partition in woodworking shop	159
5.10	Use of a screen to attenuate noise	160
5.11	A typical curve showing the drop in sound pressure level due to distance in a large rectangular factory space with low surface aborption (after Lindqvist)[8]	164
6.1	Nomogram for calculation of equivalent continuous sound level, L_{Aeq}	170
6.2	Flowchart for hearing conservation programme	172
6.3	Audiogram – slight noise-induced hearing loss	184
6.4	Audiogram – more marked noise-induced hearing loss	185
6.5	Typical headphones for audiometric tests	186
6.6	Audiometric booth for use in ambient noise levels of greater than 45 dB(A)	187
6.7	Audiogram - subject's age 43: result; category 3: referral (Lt. high freq)	189
7.1	Typical fan noise spectra	195
7.2	Theoretical value of end reflection attentuation	204
7.3	Acoustic index run of a ventilation system	204
7.4	Silencers installed in an air-conditioning system	211
7.5	Expansion plenum chamber	213
7.6	Dynamic magnification factor	219
7.7	Transmissibility	220
7.8	Shock and vibration isolator	223
7.9	Ribbed mat isolator	224
8.1	Addition of waves showing incomplete cancellation	227
8.2	Simplified schematic and block diagram of monopole active attenuator	228
8.3	Tight-coupled monopole in duct	229
8.4	Attentuation obtained using two cascaded tight-coupled monopole attentuators with high quality equipment	229
A.1	Sinusoidal oscillation in complex notation	240
A.2	Complex representation of a travelling wave	242
A.3	Displacement of air in a sound wave	244

Tables

2.1	Octave band frequencies	25

2.2	Meter averaging times	39
2.3	Crest factor requirements of current equipment standards	40
2.4	Instrument selection tolerances	42
3.1	Some units and parameters used to quantify noise exposure	49
3.2	Value of statistical parameters for traffic noise shown in Figure 3.2	54
3.3	Uses of different parameters	56
3.4	Recommended maximum L_{10} levels	58
3.5	Approximate insulation values of windows	62
4.1	Conversion factors for selected insulation ratings	97
4.2	Guide values for insulation between dwellings	99
4.3	Typical cost-performance data for remedial works	132
5.1	Sound pressure levels, dB re 20 μPa, 4m from saw	161
6.1	Showing equal noise dose values based on the UK 90 gB(A) 8-hour L_{eq}	173
6.2	Mean values of attentuation for seven hearing protection devices (after Martin)[4] evaluated and presented in accordance with BS 5108.[3] (Some of these devices may be no longer available commercially.)	176
6.3	The derivation of the assumed protection for a hearing protector	177
6.4	The deduction of the assumed protection from noise, to arrive at the SPL at the user's ear	177
6.5	The values of the arbitrary intensity units 'I' corresponding to SPLs from 60 dB to 119 dB	178
6.6	The conversion of octave band SPL at the user's ear to dB(A). The 'I' value of 0.0708 corresponds to 79 dB(A) to the nearest dB	179
6.7	Chart for categorization of hearing levels	188
7.1	Fan spectrum correction factors	196
7.2	Correction for fan efficiency	197
7.3	Sound reduction indices for common duct materials	200
7.4	Attenuation in straight sheet metal ducts	202
7.5	Attenuation provided by 90° mitred bends without turning vanes	202
7.6	Attenuation calculation for system in Figure 7.3	205
7.7	Corrections for low turbulence duct fittings	206
7.8	Corrections to overall sound power level for diffusers and grilles	207
7.9	Properties of commercial isolators	224

Introduction

The authors have attempted to provide a book which is academic in the best sense, that is the application of knowledge to the solution of real problems. Save for a minimal amount for guidance in the first Chapter and the Appendix, theoretical derivations and treatment have taken second place to a 'hands on' practical approach.

The finite size of any book is a constraint on the amount of detail that can be contained within it. This is particularly so when the emphasis is on practical applications covering a wide area, but most of the relevant topics in building and environmental acoustics are covered and in a way which provides points of ready reference for the solution of problems likely to be encountered by those concerned with noise control in the built environment. Because of this central theme the chapters are, to a significant extent, independent and may be referred to individually.

The contents of this book represent a significant part of the applied topics of the Masters Degree in Environmental Acoustics offered by the Institute of Environmental Engineering (formerly the National College of Heating, Ventilating, Refrigeration and Fan Engineering). The course is some ten years old and through the participation of consultants and others in the forefront of development, has been continuously updated to maintain its topicality and concern with current issues. Several of these have contributed to this book and ensured its usefulness to practitioners in the field. This book, then, presents discussion and review of seven key areas presently of interest to those concerned with building and environmental acoustics.

The Institute of Environmental Engineering is the major UK education and research establishment concerned with all aspects of the control of the internal environment of buildings and has received strong support from the Science and Engineering Research Council, the National Advisory Body for Public Sector Higher Education and last, but by no means least, industry. Noise is an increasingly important element in building design and the Acoustics Group, in an effort to disseminate its expertise to the largest audience, offers the various topics covered in this book as a series of individual one-day short courses.

The Editors thank all those who have assisted in producing this book, not least the authors, and hope that it will be a useful handbook to those concerned with practical noise control measures in the built environment.

John Roberts

Chapter 1

SOME THEORETICAL CONSIDERATIONS

M. L. Vuillermoz, Acoustics Group, Institute of Environmental Engineering, South Bank Polytechnic

Characteristics which single out applied acoustics from most other branches of applied science are the high levels of subjectivity which must inevitably enter into its analytical processes and the frequent need to accommodate discrepancies between theoretical prediction and actual performance.

Fortunately these characteristics tend to complement each other as the subjective response of the ear displays relatively poor resolution and so allows a reasonable degree of latitude in the achievement of final design objectives; but they do place a burden on the acoustician, who must always use personal judgement in the process of arriving at viable solutions to practical noise problems.

Rarely can an acoustic problem be overcome by the uncritical application of a mere theoretical solution. Careful appraisal of the relative importance of the parameters describing the situation and the adaptation of relevant theory will usually be needed to achieve a satisfactory outcome; this can be carried out only with a clear understanding of the fundamental physical processes involved.

In this chapter an attempt is made to provide a balanced introduction to the underlying physical principles relevant to noise control engineering. It does this by looking at the fundamentals of those topics which the authors of the text have used implicitly in their work, but in a way which emphasizes physical interpretation rather than strict mathematical rigour. Further theoretical details are developed in the Appendix.

This section does not provide a systematic introduction to theoretical acoustics as such; for this purpose the reader is referred to the many relevant texts which are available, including those listed at the end of the chapter.

Vibrating systems

With few exceptions sound is generated when a vibrating object causes compression and rarefaction of the air adjacent to its outer surfaces. Sound energy in turn when falling on a surface will tend to make that surface vibrate. An understanding of the way an object responds to an oscillating force and in particular the quantities which determine its ensuing motion is therefore important both from the point of view of the primary source of sound and for any subsequent interaction the sound might make with matter.

All objects capable of vibrating about an equilibrium position will have three properties in common; mass, a restoring force, and some form of energy loss mechanism. Although the relationships between these properties can become very complicated, the general behaviour of most oscillating systems may be obtained from an analysis of the very simplest arrangement, that of a mass hanging on the end of a spring.

Figure 1.1 illustrates such a system in equilibrium with the spring extended sufficiently to balance the gravitational force acting on the mass. The dashpot represents the energy loss process. If the mass is displaced in the vertical direction from this resting position two forces come into play, one which acts to restore the system to equilibrium and the other which acts to resist motion. The magnitudes of these forces will depend on the properties of the relevant elements of the system but for present purposes we will simply take the restoring force to be proportional to the displacement, ξ and the resistive force proportional to velocity, $d\xi/dt$.

These relationships may be expressed mathematically as

$$F_1 = -s\xi \qquad \text{restoring force}$$
$$F_2 = -r\,d\xi/dt \qquad \text{resistive force} \qquad (1.1)$$

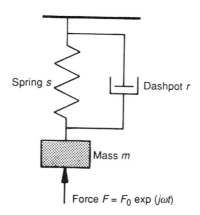

Figure 1.1   *Representation of a mass on the end of a spring*

where s is called the stiffness of the spring, and r the resistance. The negative signs are included to indicate that the forces act to oppose an increase of displacement or velocity.

Although these relationships are found to apply in a very wide range of cases it should not be assumed that they represent the only possibilities. When, for example, the spring extends beyond its elastic limit, or air resistance plays a significant role, the relationships become more complicated and the following analysis is found to be incomplete.

We now consider a sinusoidal force of angular frequency ω acting on the mass, which may be written

$$F = F_0 \cos \omega t \tag{1.2}$$

This force may be, for example, that which is exerted on the diaphragm of a loudspeaker due to variations of current in its coil winding, or that due to an acoustic pressure wave falling on a surface. Such a force will cause the mass to vibrate with a velocity u given by (see Appendix)

$$u_1 = \frac{F_1}{[r^2 + (\omega m - s/\omega)^2]^{0.5}} = \frac{F_1}{|Z_m|} \tag{1.3}$$

where u_1 and F_1 would represent simultaneously the peak or root mean square values of the relevant sinusoidal functions, and $|Z_m|$ is called the *mechanical impedance*.

The importance of this equation is that it gives the variation of the magnitude of the velocity of the mass as the frequency of the applied force is varied. A typical plot of velocity against angular frequency is given in Figure 1.2,

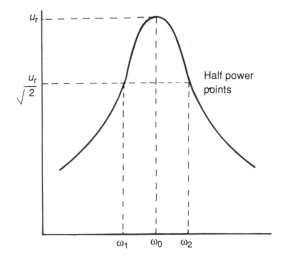

Figure 1.2 *Resonance curve showing half power points*

which shows that the velocity reaches a maximum at a particular frequency ω_0. This is the condition known as *resonance*, which comes about when

$$\omega_0 m = s/\omega_0 \tag{1.4}$$

which on substitution into equation 1.3 shows that the impedance takes its minimum value of r at ω_0.

Many of the problems encountered by the acoustician will involve either achieving resonance in a particular system or making sure that resonance does not occur, so for most purposes we will be concerned with the resonant condition itself or the system operating as far away from resonance as possible. For the latter we would be dealing with angular frequencies much larger or much smaller than ω_0, so that in either case the resistance term will be small compared to the other terms and may be ignored. For such high and low frequencies the expressions for mechanical impedance may be simplified to

$$\left. \begin{array}{ll} |Z_m| = -s/\omega & \omega \ll \omega_0 \\ |Z_m| = \omega m & \omega \gg \omega_0 \end{array} \right\} \tag{1.5}$$

This tells us that for a constant amplitude force, at low frequencies the velocity of vibration is almost entirely controlled by the stiffness of the spring and that it would be little affected by changes in the mass, and that the reverse is true at high frequencies. For any vibrating system these extremes are called the *stiffness-controlled* and the *mass-controlled regions* respectively, and for most purposes in these regions the appropriate approximate expressions for $|Z_m|$ given above may be used.

Sound power radiation from a vibrating object constitutes an additional power loss mechanism and should of course be included in the analysis with its own resistance term. For sound radiators such as the diaphragm of a loudspeaker this would normally be small compared to the internal mechanical resistance of the suspension, but when air vibrating at the end of a tube constitutes the source of sound, as for example with an open-ended organ pipe, the resistive term is almost entirely due to the power lost as radiated sound. Fortuitously the resistive force due to this effect is still proportional to velocity as was originally assumed, so that the general analysis remains valid even in this case, but as we shall see later the constant of proportionality may change with changes in the wavelength of the radiated sound.

The wave equation

A sound wave in air is essentially a sinusoidally varying displacement of the air molecules from their equilibrium position, which takes place in time and over distance. When discussing the properties of sound it is useful to be able to describe this motion in the form of an equation expressed in terms of measurable parameters and which is also mathematically convenient. Although the displacement of the molecules or particles in a sound wave is along the direction of travel, see Figure 1.3, a surface water wave provides a good analogy to what is occurring on the understanding that here the direction of displacement is perpendicular to the direction of propagation.

The mathematical representation of this type of plane travelling wave in terms of the instantaneous displacement ξ is

$$\xi = \xi_0 \exp[j(\omega t - kx)] \tag{1.6}$$

where ω is the angular frequency of the oscillating motion, as used in the preceding section, and k is a parameter known as the wave number which is 2π divided by the wavelength (x and t are distance and time respectively). As the velocity of sound c is equal to the product of frequency and wavelength, in terms of these parameters $c = \omega/k$. The reason for expressing the displacement in terms of ω and k instead of frequency and wavelength directly is that it avoids the frequent inclusion of factors of 2π in various formulae describing wave motion. The full physical interpretation of equation 1.6 is discussed in the Appendix.

Equation 1.6 describes a plane wave travelling to the right; it may be shown that if the negative sign in the formula is changed to positive it gives the equation of a plane wave travelling to the left.

The particular advantage of expressing the plane wave equation in this way is that it makes certain mathematical manipulations very easy to carry out. For example, to differentiate the function with respect to time we simply multiply it by $j\omega$, so that

$$\partial \xi / \partial t = j\omega \xi_0 \exp[j(\omega t - kx)] = j\omega \xi \tag{1.7}$$

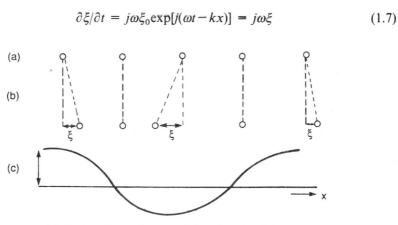

(a) Rest positions of representative air molecules
(b) Molecules displaced during passage of sound wave
(c) The displacement plotted as a function of position

Figure 1.3   *Representation of a sound wave*

Similarly, to differentiate with respect to distance we multiply by $-jk$ to give

$$\partial\xi/\partial x = -jk\xi_0\exp[j(\omega t - kx)] = -jk\xi \qquad (1.8)$$

Both these manipulations are used in later sections.

Characteristic impedance

We may now use the defining equation 1.6 to obtain the ratio between the acoustic pressure and the particle velocity in a plane wave. This ratio, which gives a measure of how the medium responds to acoustic pressure variations in it, is known as the *characteristic impedance* and is an important parameter used to quantify a number of acoustic effects. If a relatively low pressure produces a high particle velocity the ratio would be small and the medium would be described as having a low impedance. The impedance of a hard surface such as a loudspeaker diaphragm will generally be high compared to the impedance of sound waves in air.

From equation 1.6 we obtain the instantaneous velocity by straightforward differentiation of the displacement with respect to time to give

$$u = \partial\xi/\partial t = j\omega\xi \qquad (1.9)$$

and for acoustic pressure we use the equation (see Appendix)

$$p = -K(\partial\xi/\partial x) \qquad (1.10)$$

where K is the bulk modulus of air, and the derivative of displacement with respect to distance

$$\partial\xi/\partial x = -jk\xi$$

to give

$$p = jkK\xi \qquad (1.11)$$

The ratio of pressure to velocity then becomes

$$\frac{p}{u} = \frac{kK}{\omega} \qquad (1.12)$$

and as $\omega/k = c$, and $c = (K/\rho)^{0.5}$ (where ρ is the density of air, see Appendix) this ratio can be written

$$p/u = \rho c \qquad (1.13)$$

This tells us that for the plane wave the acoustic pressure and particle velocity are in phase (that is, they increase and decrease together with time), and that their ratio, the characteristic impedance of the medium for the plane wave, is determined by the density of air and the speed of sound only. This relationship will be used later and in particular when discussing the measurement of intensity.

Sound intensity

Having considered the structure of a sound wave in terms of pressure, displacement and velocity, we now turn to the question of energy content. A sound wave transports energy from one place to another, which energy, for example, when falling on the ear causes the tympanic membrane to vibrate and so stimulates our sensation of hearing. The most useful way of expressing this energy transport is by the rate at which it crosses unit area per second. This quantity is called the *intensity*, it is expressed in watts per square metre and indicates how the power is concentrated in the sound wave. An expression for the intensity may be derived by considering the way in which the energy is carried by the wave (see Appendix). This shows that the instantaneous rate of transmission of energy from one layer of air to the next in the sound wave is the product of the acoustic pressure and the particle velocity which exist at that moment and at that location.

$$I = pu \qquad (1.14)$$

As both these quantities change sinusoidally with time it may not be immediately obvious what the average intensity will be, but we remember that in a plane wave the pressure and velocity are in phase so that when the velocity becomes negative so also does the pressure. This means that the intensity is always positive and travels with the wave, and being the product of two sinusoidal functions, its average value may be shown to be the product of their root mean squares.

If we now combine equation 1.14 for intensity with equation 1.13 for the characteristic impedance of a plane wave we arrive at the important result that the intensity of such a wave may be expressed in terms of the acoustic pressure alone, that is

$$I = p^2/\rho c \qquad (1.15)$$

This equation turns out to be of considerable value when it comes to the measurement of intensity, and forms the basis of calibration for most of our sound intensity measuring instruments. It should be noted, however, that this relationship was derived from considerations of a plane wave, and although it may apply equally for spherically expanding waves it must not be used directly in reverberant fields or when standing waves are involved. For the latter cases, which are both considered in later sections, it is shown that the direct reading on the meter may not necessarily give the correct value for the sound intensity.

Sound radiation

Equation 1.14 gives the intensity of a plane sound wave as

$$I = pu$$

This represents the sound power per unit area, so that the total power W which would travel down a long pipe of cross-sectional area S, say, carrying such a wave will be

$$W = pSu \tag{1.16}$$

but from equation 1.13 $p/u = \rho c$, so that on substitution we obtain

$$W = Su^2 \rho c \tag{1.17}$$

If this is rewritten

$$W = (Su)^2 (\rho c / S) \tag{1.18}$$

we obtain an equation which may be regarded as analagous to that describing the electrical power loss in a resistor R carrying a current I, that is, $W = I^2 R$.

For the plane wave case $\rho c/S$ may be considered equivalent to a resistance and called the acoustic resistance of the pipe, and Su equivalent to a current and known as the volume velocity.

It is often convenient to consider a source of sound, which will usually be some type of vibrating surface, as if it were a generator of volume velocity. This volume velocity, the root mean square normal velocity of the source multiplied by its surface area is called the source strength, and will feed into the acoustic radiation resistance presented to it by the surrounding medium. A vibrating loudspeaker, for example, placed over the opening of an infinitely long pipe will feed acoustic power into the pipe equal to the square of its source strength multiplied by the pipe's acoustic resistance, as in equation 1.18.

For pipes of finite length some sound will be reflected from the far end and the input resistance will be changed as a result. Such changes may be used to explain pipe resonances and are considered in more detail in the next section.

When the source radiates into free space the same analogy may be applied but, not surprisingly, the resistive term becomes rather more complicated than the simple $\rho c/S$ of the plain pipe. For the general case it is found that the acoustic pressure and the particle velocity are not necessarily in phase, and that the magnitude of the resistance changes with the relative dimensions of the source and the sound wavelength.

Sources which are small compared to the wavelength of the sound they emit, whatever their shape, behave to a first approximation like a pulsating sphere. The source strength would be, as before, the mean area of the vibrating surface times the normal velocity, but because the acoustic radiation resistance in this case is $\rho c k^2/4\pi$ the power radiated would be

$$W_r = \frac{\rho c k^2}{4\pi}(Su)^2 \qquad (1.19)$$

As this power radiates uniformly from the source the intensity of the sound field a distance r away would be

$$I_r = \frac{\rho c k^2 (Su)^2}{16\pi^2 r^2} \qquad (1.20)$$

At high frequencies when the wavelength is small compared to the dimensions of a plane source the radiated sound becomes highly directional and the power radiated is the same as that from a loudspeaker feeding into a uniform pipe.

$$W_r = (Su)^2(\rho c/S) = u^2 \rho c S \qquad (1.21)$$

When two sources of equal magnitude and a distance h apart are made to pulsate in antiphase, that is, the surface of one moves forward as the other moves back, they form what is known as a *dipole*. Interaction between the two sources causes the radiated impedance at each to be reduced and the total power radiation is less than that from an equivalent single source or monopole. The total power radiated is

$$W_h = \frac{\rho c k^4 h^2}{12\pi}(Su)^2 \qquad (1.22)$$

At positions which are equidistant from the two sources the sound radiated from each will tend to cancel. This results in the overall sound field taking an 'hour glass' shape with the maximum intensities occurring in both directions along the line joining the sources. Because the total power radiated is less than that from a single source adding an antiphase component, by for example allowing sound to be radiated from the back as well as the front of a loudspeaker, causes the total sound emission to be reduced and provides the basis of an effective noise reduction technique.

Influence of the enclosure

In the cases considered above we were dealing with sound radiating into free space with no waves reflecting back on to the source. When reflection does take place it usually has a significant effect on the sound power radiated and will need to be taken into account particularly when a source operates in an enclosure.

Essentially what occurs is that a reflected wave interferes with the incident wave to change the radiation resistance at the source. This change can alter the power radiated in two ways. For the case of a resulting increase in the radiation resistance, provided the volume velocity remains steady, the output power would be expected to rise – see equation 1.18 – but the increase of resistance at the source at the same time tends to reduce the velocity and thus make the output power smaller. There are therefore two effects operating which are seen to act in opposition, and the resulting power output will depend on the relative magnitudes of the resistance of the source and the radiation resistance into which the source feeds. This is an example of a phenomenon which occurs whenever energy is transferred from one system to another, and in general it is found that maximum power is transmitted when the resistance of the load is equal to that of the source.

In building acoustics we would be concerned chiefly with two types of source, the high resistance type such as the diaphragm of a loudspeaker or other hard vibrating surface directly coupled to a receiver, and the low resistance case where there is an air gap between the source and receiver. A receiver here refers to a room, a length of ducting or any other type of cavity. The condition for maximum power transfer, the condition known as resonance, will occur when the resistances are equal, in which case the source and receiver are said to be matched.

To illustrate this whole phenomenon in a simple way we may consider the case of a loudspeaker feeding sound into a finite length of tube. If the tube is terminated by a hard surface, there can be little or no movement of the air particles at the termination and so a reflected wave must be created at that point with its velocity amplitude acting to cancel that of the incident wave. Because pressure and velocity are in phase and because the incident and the reflected waves are travelling in opposite directions this will result in the acoustic pressure amplitude at the termination being almost doubled. This high pressure coupled with the low velocity means that the acoustic resistance at the termination is high as would be expected.

What is significant here is that because of the relationship between incident and reflected waves the same state of affairs will occur within the pipe at a distance of half a wavelength from the termination, and at that position the pressures will add and the velocities subtract to produce another region of high resistance. This will be repeated at half-wavelength intervals along the tube. By the same token, between these high resistance regions there will be regions of low resistance where the pressures subtract and the velocities add. The resistance acting on the source will therefore depend on where it is located with respect to this pattern.

Positions of minima and maxima are called *nodes* and *antinodes* respectively, and it is seen that pressure and velocity nodes do not coincide but occur one quarter of a wavelength apart. The type of wave pattern described is known as a *standing wave*.

Resonance for the system will then depend on the way that the source is connected; if the speaker is attached directly to the open end of the tube the source and receiver will be matched when the input resistance of the tube is high, that is, when it is an even number of quarter wavelengths long; but if the speaker is away from the open end and relies on the intervening air to transport sound to it, its effective resistance will be low so that resonance will occur when the input resistance is also low, that is, when the tube is an odd number of quarter wavelengths long. Similar reasoning may also be used to obtain the equivalent resonant conditions for an open terminated tube.

Sound radiation into a room

The effect on the source of sound reflection described in the previous section will apply equally to the more general case of a source operating in a room. Here the reflections will be more complicated and their influence on the source more difficult to describe, but the overall effect turns out to be similar and of importance particularly regarding the acoustic quality of auditoria.

If we consider sound in a three-dimensional enclosure such as a room it will be seen to be continuously striking the walls and being reflected back into the main field. At each encounter some sound energy will be lost so that eventually the sound would die away unless it were being replenished. The process will generally take place in a random fashion so that at any position in the enclosure sound will be arriving from all directions.

As the walls of the room would usually be hard, when the sound waves fall on them the reflected wave will be of such magnitude and phase that the normal particle velocities cancel and the acoustic pressures add as in the previously considered case of the terminated tube. The wave resistance at the walls will therefore be high. What occurs in the space between may be described by considering a source such as a loudspeaker placed between two parallel walls or between the floor and ceiling. If the source emits sound of one frequency the reflected waves from each surface will, with the incident wave, start to form two independent standing wave patterns, but the two patterns would be of arbitrary phase and their effects at the speaker will tend to even out with time. This situation changes if the frequency of the source is such that the distance between the reflecting surfaces is an integral number of half wavelengths. In this case the two standing wave patterns come into phase and standing pressure and velocity nodes are formed at half-wavelength intervals between the surfaces. The source may then find itself located at a position where the wave resistance is high, that is, at or near a pressure antinode, so that it becomes better matched to the sound field, and more power can be fed in to sustain or increase it.

The effects of reflections from other walls may be taken into account by considering the components of the wavelengths in the direction of each axis and imposing the condition that standing waves will form only when an integral number of half-wave components fit between the opposite walls.

For a rectangular room of sides l_x, l_y, and height l_z this leads to the result that the frequencies at which three-dimensional standing wave patterns can form are given by

$$ f = \frac{c}{2}\left[\left(\frac{n_x}{l_x}\right)^2 + \left(\frac{n_y}{l_y}\right)^2 + \left(\frac{n_z}{l_z}\right)^2 \right]^{0.5} \tag{1.23} $$

where n_x, n_y, and n_z can be any integers, 0, 1, 2 ...

These frequencies are known as the *eigentones* of the room, and the wave patterns the normal room modes. A particular pattern or mode may be characterized by the three integers (in the order *x, y, z*) which describe it, so, for example, the [1, 1, 0] mode will be a standing wave between the walls of the room with components parallel to the walls each of wavelength equal to twice the wall length, but with no pressure variation between floor and ceiling.

At low frequencies the modes are relatively widely spaced and become closer as frequency increases. They represent the frequencies at which power from a source is likely to be enhanced due to favourable matching, but this will also depend on the positioning of the source at the correct location. A high resistance source at the centre of a room, for example, is unlikely to generate the lowest eigentone as the resistance for that mode will be low at that particular position. To excite the maximum number of eigentones the source would need to be located at a corner of the room as here the resistance will be high for all the room modes.

The formation of standing waves in an enclosure means that to measure accurately the mean sound intensity in a room the sound level meters should be placed in positions away from the walls to avoid those regions where the sound pressure is increased due to reflections, and which might otherwise lead to too high a reading.

An important consequence of the effect of the room modes is that any sound produced in the room will tend to have these frequencies enhanced. This enhancement, which is determined largely by the room dimensions, becomes to the listener a natural result of hearing the sound in that room. It is therefore regarded as desirable that any alterations, say, aimed at improving the acoustics of a room should not disturb these naturally occurring resonances or else the quality of listening could be impaired.

Absorption of sound in a room

The formation of standing waves in an enclosure considered above is of course a consequence of the wave nature of sound. When we wish to find the way sound energy is absorbed in a room it becomes more convenient to treat the sound as if it travelled as rays and to ignore its wave properties. This is made possible because, as we saw, unless the frequencies are very low, the room modes are closely spaced so that the energy is more or less evenly spread in the enclosure and would be travelling uniformly in all directions. This does impose a limit on the following results which should not be assumed to apply at those frequencies which make the wavelength comparable with the room dimensions, or when the absorption is very large or when there are large differences in the absorption characteristics of different surfaces.

Consider a source feeding sound energy into a room. To start with the energy content of the space will increase but as the walls will always absorb a certain fraction of the sound power falling on them an equilibrium state must arise when the power being lost from the sound field becomes equal to the power W radiated by the source.

If the intensity of the sound falling on a particular surface is I, the power absorbed per unit area may be written αI, where α is called the coefficient of absorption. This quantity, which may have a value between 0 and 1, representing respectively no absorption and total absorption, is a property of that surface and generally varies with frequency.

The equilibrium state, which is when the total power emitted is equal to the total power absorbed, may be written

$$W = \sum_n S_n \alpha_n I = AI \qquad (1.24)$$

where S_n and α_n are the area and absorption coefficient respectively of each surface. The effects of all the surfaces have been summed with I taken to be the same for all surfaces. The quantity A is the total absorption.

This formula turns out to be of considerable importance in building acoustics, and forms the basis of many calculations which the acoustician will need to perform. It is important to note that the intensity I referred to here is not the intensity which would be indicated on a normal sound level meter placed in the room. As explained previously such meters are calibrated on the assumption that the sound waves falling on them are either plane or spherical. For random incidence, which is the situation here, these calibration assumptions need to be changed as shown below.

On the assumption that we may treat the energy as uniformly spread throughout the room, there will be at any instant a certain energy density ε associated with the sound field. The energy content of a small volume dV will then be εdV, which will, as time passes, spread out radially at the speed of sound to fall eventually on to the room surfaces.

The rate at which the energy from all parts of the room falls on to an area of the surface can be calculated and from this we may obtain an expression for the intensity of such a field of

$$I_r = \varepsilon c/4 = p^2/4\rho c \qquad \text{[reverberant field] (1.25)}$$

This may be compared with the equivalent expression for the intensity of a plane wave falling on a surface which is

$$I_p = \varepsilon c = p^2/\rho c \qquad \text{[plane wave] (1.26)}$$

So we see that instruments calibrated for use with plane waves could give a reading which is in error by a factor of four if they are assumed to provide the intensity on the inner surface of an enclosure from a pressure measurement alone. It is important therefore to distinguish between the sound pressure level, L_p, and the intensity level, L_I, which are the equivalent quantities expressed in dB (see the final section to this chapter) when dealing with reverberant fields. For plane and spherical waves L_p and L_I are numerically equivalent, but in reverberant fields L_I for the surface and L_p are related by

$$L_I = L_p - 6 \qquad \text{dB} \qquad (1.27)$$

Sound decay and reverberation

We may use the results derived in the last section to obtain the way sound decays in a room once the source has stopped. The time for sound energy in an enclosure to decay by a factor of one million is called the *reverberation time*, and this turns out to be an extremely useful parameter for describing the acoustic properties of rooms in general.

With the source silenced the instantaneous rate of loss of total energy in the interior space of the room will equal the rate of absorption of energy by the walls. Using equations 1.24 and 1.25 we may write

$$V(\mathrm{d}\varepsilon/\mathrm{d}t) - Ac\varepsilon/4 = 0$$

or

$$\mathrm{d}\varepsilon/\mathrm{d}t - Ac\varepsilon/4V = 0 \qquad (1.28)$$

The solution to this equation, which may be verified by substitution, is

$$\varepsilon_t = \varepsilon_0 \exp(-Act/4V) \qquad (1.29)$$

which shows that the energy density falls exponentially with time as might be expected. Because the intensity is proportional to the energy density the reverberation time will be the time T taken for the ratio $\varepsilon_t/\varepsilon_0$ to become 10^{-6}, that is

$$\exp(-AcT/4V) = 10^{-6} \tag{1.30}$$

If we take the common logarithm of both sides of this equation we obtain

$$(AcT/4V)\lg(\exp) = 6 \tag{1.31}$$

and on substituting the appropriate constants we arrive at the Sabine formula for reverberation time

$$T = \frac{0.161V}{A} = \frac{0.16V}{\sum a_n S_n} \tag{1.32}$$

While this equation may be used in cases where the mean absorption of the walls is low, it becomes inaccurate when the absorption is high. Indeed if the walls absorb all the sound energy falling on them, that is, when α is equal to one, this formula gives a finite value for T instead of the expected zero, showing that it must be incomplete.

If we do not treat the absorption as an overall effect but instead consider individual reflections we arrive at a formula for reverberation time which overcomes this difficulty. Let the initial intensity of sound incident on a surface be I_0, then after one reflection its intensity will be $I_0(1-\alpha)$, and after n reflections $I_0(1-\alpha)^n$.

The number of reflections taking place in a time t will be the total distance travelled by the sound in that time, ct, divided by the mean distance between reflections. This latter may be shown to be $4V/S$, so the intensity after a time t will be

$$I_t = I_0(1-\alpha)^{cSt/4V} \tag{1.33}$$

If as before we substitute for $I_t/I_0 = 10^{-6}$ when $t = T$, and take logarithms of both sides

$$T = \frac{0.161V}{-S\ln(1-\alpha)} \tag{1.34}$$

This is the Norris-Eyring formula for reverberation time which because $-\ln(1-\alpha) = \alpha$ when α is small becomes the same as the Sabine formula when absorption is low. At the other extreme when $\alpha = 1$, $-\ln(0) = \infty$ so the formula gives the correct value for T in this case. It should be noted that the limitations listed for the analysis given in the last section apply equally to the case of reverberation time.

The importance of reverberation time is that it provides a quantity which

can be easily obtained experimentally and which gives a good measure of
the sound-absorbing properties of a room. From a knowledge of the total
room absorption we may estimate the magnitude of the reverberant sound
field for any source of known output power, using formulae based on equation
1.24. Equally the sound power emission from a source may be obtained from
the magnitude of the reverberant field it produces in a room of known
reverberation time.

Sound transmission through barriers

In the last section we were concerned with the energy being removed from
the sound field in an enclosure by absorption at the walls. We now turn
our attention to the way sound energy falling on a wall or indeed any other
type of barrier may be transmitted through it to emerge on the far side.
Such barriers may be made up of a single layer of homogeneous material
or be of multilayered construction. In the following analysis we consider first
the single barrier and see later how the model which emerges may be modified
for application to the case of a double barrier. For more complicated barriers
the theory becomes intractable and their performance may be best obtained
from empirical measurement. The analysis is simplified and made more readily
understandable by considering the essential steps first and making any
necessary approximations at an early stage.

The types of barrier most likely to be of interest are walls of a dwelling,
window glass or plasterboard partitions where at medium frequencies the wavelength of sound is much larger than the barrier thickness. In this case, except
at the highest frequencies, the front and back surfaces of the barrier will
move in phase when the sound passes through it.

A plane sound wave falling on the barrier at an angle θ will generate a
reflected wave on the same side, in the way described above and a transmitted
wave which will pass through to the far side. The sequence of events leading
to this transmission is an acoustic pressure build-up on the first surface due
to a combination of the incident and reflected waves which causes the barrier
to vibrate and which in turn generates a sound wave in the air adjacent
to the second surface, as shown in Figure 1.4.

The barrier may be treated in the same way as any other vibrating system
and, as previously shown, the velocity obtained from the applied force divided
by the mechanical impedance as given in equation 1.3. If we take unit area
of the barrier this force will be simply the pressure difference across it, so
that if Z_m is the impedance for unit surface area, we may write for the normal
velocity u.

$$u = (p_i + p_r - p_t)/Z_m \qquad (1.35)$$

Here p is the pressure amplitude at the barrier ($x = 0$), and subscripts i, r,
and t refer to the incident, reflected and transmitted waves respectively.

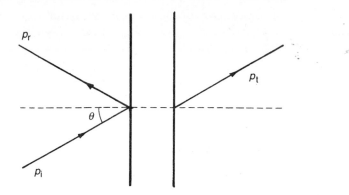

Figure 1.4 *Sound transmission through a barrier*

We now make the assumption that the air on both sides of the barrier remains always in contact with the barrier surfaces. This results in the combined normal velocities of the incident and reflected waves being the same as the normal velocity of the barrier and the same as the normal velocity of the transmitted wave. This is known as the *continuity of velocity* across the surfaces and although it does not necessarily apply at very high intensities it will certainly do so at the intensities likely to be encountered in the normal course of events.

Expressed mathematically this becomes

$$u_i\cos\theta + u_r\cos\theta = u = u_t\cos\theta \qquad (1.36)$$

But for plane waves $p_i/u_i = \rho c$, $p_r/u_r = -\rho c$, and $p_t/u_t = \rho c$, and if these are substituted into the above equation we obtain

$$p_i - p_r = cu/\cos\theta = p_t \qquad (1.37)$$

Combining this equation with equation 1.35 and after some manipulation we obtain the important result that

$$\frac{p_i}{p_t} = 1 + \frac{Z_m\cos\theta}{= 2\rho c} \qquad (1.38)$$

This gives the relationship between the incident and transmitted acoustic pressure waves, and because the energy content of a wave is proportional to the square of the acoustic pressure it can provide the ratio of the incident to the transmitted energy at the angle θ.

Sound transmission through most barriers will generally be the result of sound falling on them randomly from all directions, we therefore need to combine the effects of sound incident at all angles of θ.

This results in an expression for the ratio of the sound transmitted through the barrier to the sound incident upon it, a quantity known as the sound transmission coefficient τ, given by,

$$\tau = (2\rho c/|Z_m|)^2 \ln[1 + (|Z_m|/2\rho c)^2] \qquad (1.39)$$

The final form of this transmission equation will depend on the appropriate expression to be inserted for $|Z_m|$ (see equation 1.5). There is generally no single resonant frequency for a barrier as it may experience many vibrational modes as frequency changes, so we cannot simply treat it as stiffness controlled at low frequencies and put $|Z_m| = s/\omega$. At higher frequencies on the other hand it does become possible to consider the barrier as mass controlled and the replacement of $|Z_m|$ by ωm in equation 1.39, where m is the mass per unit area, will give a reasonable estimate for the transmission ratio for a wide range of barrier types and over a range of frequencies. In this frequency range the barrier is said to obey the *mass law*.

At frequencies above the mass law region a phenomenon occurs known as the *coincidence effect*, which is associated with flexural waves in the barrier and results in a marked increase in the power transmitted through it. Natural flexural waves which may occur in any sheet of material, travel with a velocity determined by the dimensions, the properties of the material and the driving frequency.

Sound waves which are incident on the surface of a barrier at an angle θ to the normal will drive flexural waves across the barrier as points of constant phase sweep over it. The velocity of these waves c_f will be given by $c_f = c/\sin\theta$. A driven flexural wave will generally be of small amplitude except at such a frequency and angle when its velocity coincides with that of a natural flexural wave. When this occurs the impedance of the barrier becomes very small and the fraction of the incident sound transmitted through it becomes much larger; see Figure 1.5.

The lowest frequency at which this coincidence effect occurs takes place when the incident sound is at grazing incidence, ($\theta = 90°$). This frequency is called the critical frequency f_c and marks the point from which the sound transmission starts to increase. The critical frequency for most materials is given by

$$f_c = (c/2\pi)(12m/Eh^3)^{0.5} \qquad (1.40)$$

where E is the Young modulus for the material and h is the barrier thickness.

The treatment of the double barrier follows the same pattern as above but with the additional effects due to reflections from the second layer taken into account. This results in an expression for the transmission coefficient which shows that at a certain frequency a type of resonance occurs, when almost all the incident sound is transmitted. The condition varies with the angle of incidence so that transmission will increase slowly to a maximum

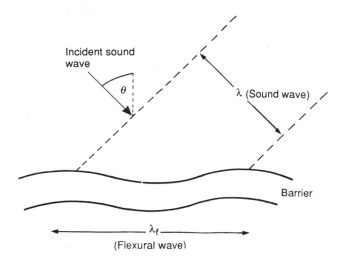

Figure 1.5   *Flexural waves and the coincidence effect*

as frequency increases, after which it falls, but this increase can reach a value when transmission is actually greater than that through a single panel of similar dimensions.

It must be clearly stated in conclusion that the models for both the single and double panels can be applied only if the panel surfaces are smooth and free from holes. The expressions for transmission become badly in error if the panel surfaces are rough or sound absorbing to any marked extent.

Working formulae and the decibel notation

Those formulae which are most frequently used by acousticians to quantify acoustic effects will be expressed in decibels (dB). The decibel, which was originally devised by electrical engineers to describe the power gain of amplifying circuits, is a logarithmic ratio unit. It is used in acoustics mainly because it provides a convenient way of analysing acoustic phenomena in a manner which broadly corresponds to the subjective response of the ear.

Formulae written in decibel notation will generally look rather different from the equivalent expressions written in linear units, and it is not always obvious how they are related. For this reason in this final section of this chapter we examine two of the more important practical formulae used in noise control engineering to show, by way of example, how they may be derived from their original defining equations.

We need first to consider the decibel as an acoustic unit. The decibel is strictly a unit for describing a power ratio. If W_1 and W_2 represent two powers, then the level difference L between them is said to be

$$L = 10\lg(W_1/W_2) \qquad \text{dB} \tag{1.41}$$

If one of the powers is specified, for example, $W_2 = 1$ pW, then W is said to have a power level L_W given by

$$L_W = 10\lg(W_1/10^{-12}) \qquad \text{dB re 1 pW} \tag{1.42}$$

Sound intensity, which is power per unit area, may also be expressed in the same way, so that an intensity of I W/m^2, would have an intensity level with respect to a reference intensity I_0 of

$$L_I = 10\lg(I/I_0) \qquad \text{dB re } I_0 \tag{1.43}$$

For plane and spherical sound waves it has been shown that intensity is proportional to the square of the acoustic pressure, so for such waves equation 1.43 may be rewritten in terms of a pressure level L_p as

$$L_p = 20\lg(p/p_0) \qquad \text{dB re } p_0 \tag{1.44}$$

The reference pressure p_0 is frequently taken as $p_0 = 20$ μPa, which pressure corresponds approximately to the threshold of hearing. In this way the decibel can be treated as an absolute measure of intensity. It is regarded as bad practice to assume the reference, which should always be explicitly stated whenever the sound pressure level is meant to represent an absolute quantity.

It should be noted that while for plane and spherical waves the sound intensity level and the sound pressure level are numerically equal when compared to the same equivalent reference, this is not necessarily the case for other types of sound field – see equation 1.27.

As our first example we take the case of a sound source in a room or other enclosure and wish to find the sound pressure level in an adjoining room due to sound transmission through a dividing partition of area S.

If W_2 represents the power transmitted through the partition, and I_2 is the sound intensity in the second room, equilibrium will be established when

$$W_2 = I_2 \sum a_n S_n = I_2 A \tag{1.45}$$

where A is the total absorption in that room – see equation 1.24.

But W_2 will be that fraction of the sound power falling on the partition in the first room which penetrates to the second. We may then write $W_2 = \tau W_1$, where τ is the sound transmission coefficient for the dividing partition and W_1 is the incident sound power.

If the sound fields in both rooms are taken to be reverberant we may write (see equation 1.25)

$$W_1 = I_1 S = p_1^2 S/4\rho c$$

$$(1.46)$$

and

$$I_2 = p_2^2/4\rho c \qquad (1.47)$$

Here p_1 and p_2 are the acoustic pressures corresponding to W_1 and W_2 respectively.

The condition for equilibrium will then be given by

$$\frac{p_1^2 \tau S}{4\rho c} = \frac{p_2^2 A}{4\rho c} \qquad (1.48)$$

Dividing both sides by the square of the reference pressure (p_0 = 20 μPa) and taking ten times the logarithm of each side we arrive at

$$10\lg(p_2/p_0)^2 = 10\lg(p_1/p_0)^2 + 10\lg\tau + 10\lg S - 10\lg A \qquad (1.49)$$

which becomes

$$L_2 = L_1 + 10\lg\tau + 10\lg S - 10\lg A \qquad (1.50)$$

The quantity $10\lg(1/\tau) = -R$ is called the *sound reduction index*, and may be put into the equation so that it finally becomes

$$L_2 = L_1 - R + 10\lg S - 10\lg A \qquad (1.51)$$

Our second example concerns the sound reduction index R of a partition, but in terms of the relevant physical properties which determine the resulting sound transmission. Equation 1.39 gives the transmission coefficient for a single partition, which with the appropriate substitution for the impedance corresponding to the mass law region becomes

$$\tau = (2\rho c/\omega m)^2 \ln[1 + (\omega m/2\rho c)^2] \qquad (1.52)$$

Again we take ten times the logarithm of both sides, and with some manipulation the equation becomes

$$R = 10\lg(1/\tau) = 20\lg(\omega m/2\rho c) - 10\lg(\ln[1 + (\omega m/2\rho c)^2]) \qquad (1.53)$$

This equation shows that if the mass per unit area of the wall is, say, doubled, the sound reduction index will increase by approximately $20\lg 2 = 6$ dB, and in the same way a doubling of frequency leads also to an increase of 6 dB. Actual partitions do not necessarily follow this rule exactly but it gives a reasonably good indication for order of magnitude calculations.

Further reading

Beranek, L. L. (1954) *Acoustics*, McGraw-Hill, New York.

Dowling, A. P. and Ffowcs Williams, J. E. (1983) *Sound and Sources of Sound*, Ellis Horwood, Chichester.

Kinsler, L. E., Frey, A. R., Coppens, A. B. and Sanders, J. V. (1982) *Fundamentals of Acoustics*, John Wiley, Chichester.

Pierce, A. D. (1981) *Acoustics*, McGraw-Hill, New York.

Chapter 2

UNITS, INSTRUMENTATION AND MEASUREMENT

K. Scannell, Lucas CEL Instruments Ltd. and **P. Colgrave**, Acoustics Group, Institute of Environmental Engineering, South Bank Polytechnic

Units and descriptors

The most common measure used to describe acoustic phenomena is sound pressure level, and it is the determination of this quantity which occupies the main part of this chapter. Sound pressure level is usually expressed in decibels and direct readings can be obtained using a sound level meter, a device that converts the pressure fluctuations associated with a sound wave into a voltage which is then displayed on a calibrated scale. Other measures of interest are intensity level and the total sound power radiated from the source. These quantities have already been defined in the previous chapter.

When these quantities are used to represent likely human response, they must be suitably adapted to allow for subjective perception, and the introduction of weighting scales into the measuring device has been one attempt to model the complex response of the mechanism of hearing.

Weightings

The ear does not exhibit a linear response to sound with regard to either frequency or level, and this means that people do not recognize equal amounts of sound energy at different frequencies as being equally loud. The ear is most sensitive to sounds in the 1–5 kHz frequency range and much less sensitive at extremely high and low frequencies.

The response of the ear to frequency and level is shown in the equal loudness level contours of Figure 2.1. The curves were established by testing a large number of subjects who listened to pure tones at different frequencies and

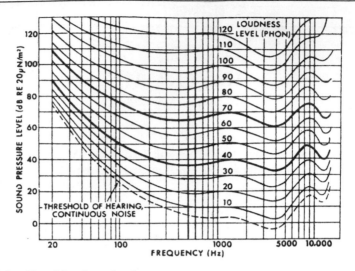

Figure 2.1   *Equal loudness level contours*
(Reproduced from Section A1 of the CIBSE Guide, by permission of the Chartered Institution of Building Services Engineers.)

decided at which sound pressure levels two sounds appeared equally loud. A line joining points of equal loudness across the frequency range is called an *equal loudness contour*. Each contour is described by the term 'phon', the unit of loudness level, and given a number equal to its sound pressure level at 1 kHz. For example, the 40 phon equal loudness contour joins the points across the spectrum judged to sound equally loud as 40 dB re 20 μPa at 1 kHz.

Many attempts have been made to develop a simple means of relating objective sound level measurements to human subjective response. The introduction of the 'A', 'B' and 'C' weighting networks into measuring devices was an attempt to imitate the ear's frequency response at different levels of sound. The weighting networks consist of filters which attenuate the sound energy at frequencies where the ear is less responsive and slightly amplify it at frequencies where the ear is most responsive. The 'A' weighting network is based on the 40 phon equal loudness contour and therefore approximates the ear's response at low sound pressure levels, the 'B' and 'C' weighting networks follow the 70 and 100 phon loudness contours respectively, approximating the ears response to higher sound levels. Today, only the 'A' weighting network is widely used as the 'B' and 'C' networks have not stood the test of time. Currently a 'D' weighting network is also occasionally used for aircraft noise measurements.

Frequency bands

Most sounds and noises that the ear detects consist of a number of different frequencies and levels. Since the ear is not equally sensitive across the frequency range to sounds of similar intensity, it is usually necessary to know the frequency content of a sound signal in order, for example, to solve a noise problem or predict likely annoyance. Although the frequency range of interest to the noise and vibration control engineer may be from dc to 100 kHz or more, the audible range is smaller than this and is usually considered to cover the frequencies from 20 Hz to 20 kHz. Sound at frequencies below 20 Hz is termed *infrasound* and, above 20 kHz, *ultrasound.*

For convenience of analysis, the audible frequency spectrum is divided into standard octave bands, the upper limiting frequency of the band being twice the lower limiting frequency. Each band is described by its geometric mean (centre) frequency. This centre frequency f_c is given by

$$f_c = (f_u \times f_l)^{0.5}$$

where f_u = upper band limit, and f_l = lower band limit.

Recommended centre frequencies are given in ISO 266,[1] and the band limits of the eight octave bands in common use are shown in Table 2.1.

In many cases the use of octave bands does not provide enough detail from which to reach adequate conclusions. Each octave band may then be subdivided to give three corresponding one-third octave bands. This provides twenty-seven bands covering the audible frequency range, to enable a much more detailed frequency analysis of the noise in question.

For scientific and experimental work where it may be essential to identify a particular individual frequency component, narrow band analysers are available, although they are generally relatively expensive. Narrow band analysers, such as the Fast Fourier Transform (FFT) type, analyse a signal over a constant

Table 2.1
Octave band frequencies

Octave band centre frequency, Hz	Approximate band limits, Hz
32	23 to 44
63	44 to 89
125	89 to 177
250	177 to 354
500	354 to 707
1k	707 to 1 414
2k	1 414 to 2 828
4k	2 828 to 5 656
8k	5 656 to 11 312

bandwidth, as opposed to the constant percentage bandwidths of octave and one-third octave filters. For example, the 0 Hz to 20 kHz range may be divided by 400 or 800 points yielding a resolution of 50 Hz or 25 Hz respectively. If the range is restricted to an upper limit of 1 kHz and 800 points are used, this would result in a resolution of 1.25 Hz.

Descriptors

Equivalent continuous sound pressure level, L_{eq}

If a sound level is reasonably constant with time, the sound pressure level may be measured via the slow weighted rms detector of a sound level meter. However, in practice, sound levels may vary widely over the period that the noise is measured and it is necessary to find a way of presenting the data in a more rational form. One increasingly popular measure is the equivalent continuous sound pressure level, L_{eq}, a time averaged level taken over a specified time period which gives a measure of the average sound energy over that period. The L_{eq} is given by

$$L_{eq} = 20 \lg 1/T \int_0^T p/p_0 \, dt \qquad dB$$

where T = the measurement period, seconds
$\quad\ \ p$ = the instantaneous sound pressure, Pa
$\quad\ \ p_0$ = the reference, pressure, 20 μPa

Frequently it is desirable to use a time averaged value of the 'A' weighted sound pressure, or the equivalent continuous sound level, L_{Aeq}.

$$L_{Aeq} = 10 \lg 1/T \int_0^T p_A^2/p_0^2 \, dt \qquad dB$$

where p_A is the 'A' weighted sound pressure and the other symbols their given meaning. L_{Aeq} is believed to give a good indication of potential risk of hearing damage and is also used to assess potential annoyance in some circumstances. In the UK the fast time weighting is used for the measurement of L_{Aeq}. If T is equal to eight hours, then L_{Aeq} can be regarded as a measure of noise dose in one working day.

Sound exposure level, L_{Ax}

When measuring discrete events such as aircraft overflights, difficulties arise in knowing when to start and stop the averaging and how to compare the measurements obtained. For example, it may be required to know which event contains the most sound energy, a five-second L_{Aeq} of 92 dB or a three-second L_{Aeq} of 95 dB. The sound exposure level (SEL), sometimes known as the *single event noise exposure level* (L_{AX}), provides a measure indicating the total sound energy of a discrete event, and is obtained by adding a time adjustment to the L_{Aeq} value. The time adjustment is given by $10\lg T/T_0$ where T is the measured time of the event and T_0 is the reference time of one second. Hence the sound exposure level is given by

$$L_{AX} = L_{Aeq} + 10\lg T/T_0 \quad \text{dB}$$

The symbol for sound exposure level is L_{AE} or L_{AX} in the case of 'A' weighted levels, but other weightings may be used if specified.

Returning to the example above, a five-second L_{Aeq} of 92 dB gives

$$L_{AE} = 92 + 10\lg 5/1 \text{dB} = 99 \text{ dB}$$

and a three-second L_{Aeq} of 95 dB gives

$$L_{AE} = 95 + 10\lg 3/1 \text{dB} = 100 \text{ dB}$$

thus the shorter signal contains more sound energy. Many sound level meters include the sound exposure level function, but since the SEL is usually measured for a discrete event, some means is necessary to initiate and terminate the integration of noise. To do this an adjustable threshold sound level is included, which may be altered in steps of either 1 or 10 dB depending on the particular measuring instrument.

Percentile sound level, L_n

To assess community annoyance from noise, long-term measurements are required, extending over many hours or even days. Since the variation in sound level can be large, it has been found convenient to describe the sound level statistically with time. A commonly used descriptor is the percentile exceeded sound level, L_n, where the subscript n refers to the percentage of the total measurement time for which that level is exceeded. For example, an L_{10} of 90 dB indicates that a level of 90 dB was exceeded for 10 per cent of the total measurement period.

The L_n may also be used in conjunction with 'A' weighting to give the percentile exceeded 'A' weighted sound level, L_{An}. L_{A90} is the 'A' weighted sound level which has been exceeded for 90 per cent of the measurement period, and in the UK is frequently taken as the background noise level. Figure 2.2 illustrates L_n. Where the use of the 'A' weighting is understood as in the measurement of traffic or environment noise it is common practice to simply write L_n or L_{eq} rather than L_{An} or L_{Aeq}.

Acoustic measurements

Sound pressure level

Due to its portability, relative low cost and ease of use, the most popular

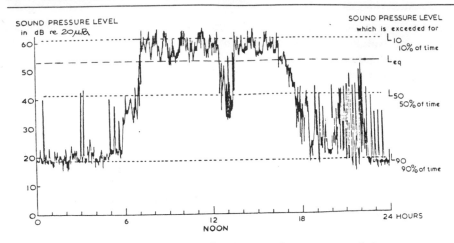

Figure 2.2　*Sound levels exceeded for the stated percentage of the measurement period (L_n)*

$$Figure 2.2 \quad Sound\ levels\ exceeded\ for\ the\ stated\ percentage\ of\ the\ measurement\ period\ (L_n)$$

(Reproduced by permission of Lucas Industrial Noise Centre.)

system for converting the transduced signal generated by the acoustic pressure into a suitable output is the sound level meter.

The sound level meter is the most basic of all acoustic instrumentation packages, although there are some fairly advanced meters on the market. Essentially a sound level meter consists of four parts: the detector, voltage amplifier, signal processor and the output. Detectors will be discussed later in this chapter but will be either a microphone or accelerometer coupled to the signal processing part of the sound level meter.

The objective of a sound level meter is to provide a visual display of the quantity to be measured, which may be either acoustic pressure or the acceleration of a surface. The required output is displayed on a decibel scale which will be calibrated according to the measured quantity. For example, if the detected quantity is acoustic pressure, the output would be calibrated in terms of sound pressure level. Almost all meters will have frequency weighted outputs intended to correlate with the human perception of the measured sound. The most common of these is the 'A' weighting, which is based on the 40 phon equal loudness contour.

Most meters currently available contain a fairly wide range of facilities, such as L_{Aeq}, L_n, and so on. Some manufacturers produce a basic meter containing the microphone, amplifiers, signal processors and digital display, but with the capacity to perform many other acoustic calculations if the necessary software is provided. This software may be supplied in the form of plug-in modules.

The majority of sound level meters normally have more than one output. Most meters will have an ac output (analogue) which allows the measured acoustic pressure to be recorded on to magnetic tape for future analysis.

A few meters actually contain a solid state memory for the same purpose. A meter of this type will need an RS232 port or an IEEE interface so that the meter can be coupled to a computer, printer, or other item of acoustic instrumentation. Figure 2.3 is an illustration of a precision sound level meter.

User calibration

The measuring system should be calibrated using a pistonphone or electro-acoustic calibrator of which the calibration level, frequency and coupler 'K' factor are known. For example, if an electroacoustic calibrator generates a level of 114 dB at 1 kHz, with a coupler having a 'K' factor of −1 dB, the microphone should be calibrated at 113 dB using the linear or 'A' weighting of the instrument. If the calibrator generates a frequency other than 1 kHz the linear setting only must be used. The fast time weighting on the meter should be used, not peak, as this will give a reading 3 dB higher.

Both the calibrator and the coupler have internal 'O' rings to ensure a tight seal and avoid acoustic leaks. The coupler should be pushed gently over the microphone ensuring that the microphone is fully up to the shoulder. The calibrator should then be pushed gently over the coupler, turned on, and the instrument display adjusted to the correct level. After the required measurements have been taken, the calibration procedure should be repeated.

Pre-measurement checks

Before taking instruments out for a sound survey it is necessary to follow a rigorous checking procedure to ensure that accurate and reproducible results are obtained. The first, but most frequently overlooked, procedure is a visual check of the equipment. This stage is particularly important if the instrument is shared. Some of the more obvious questions that should be asked are:

1   Is the body of the instrument damaged?
2   Are there any signs that it has been dropped or otherwise mistreated?
3   Is the microphone in good condition?
4   Is the windshield clean?
5   Is there any sign of leakage in the battery compartment?

When the visual check is satisfactorily completed, the microphone, preamplifier and sound level meter may be carefully assembled where necessary. The batteries may then be fitted and the meter used to check that they are adequately charged. The battery life and minimum voltage will always be stated in the manufacturer's literature. The battery level should be frequently checked during the measurement period and if the battery voltage fails all the measurements since the previous battery check should be discarded. At least one set of spare batteries should be taken on surveys, and the equipment should be recalibrated if a battery change is necessary. Discharged batteries should not

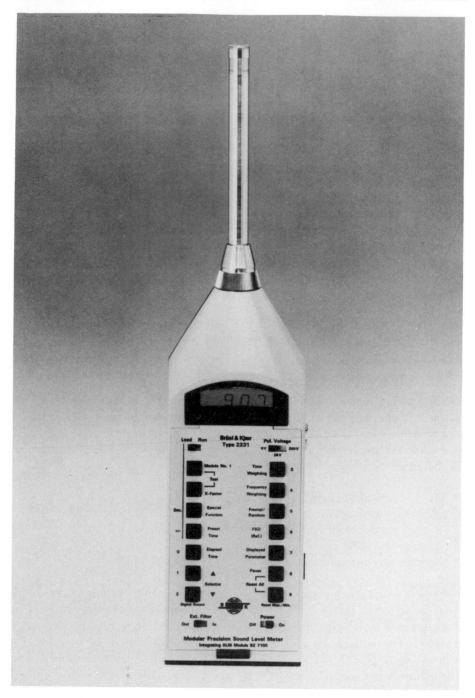

Figure 2.3   *Precision sound level meter*
(Reproduced by permission of Bruel and Kjaer (UK) Ltd.)

be left in either the instrument or its case since battery acid may cause extensive damage.

Additional test procedures

1 Testing between range linearity To test the linearity with level of the instrument, the measurement range may be varied while the calibrator is switched on. Against a reference range (for example 70 dB) the difference in sound pressure level for different measurement ranges must not vary by more than 0.9 dB to be within IEC specification.

2 Testing within range linearity This is done by using two calibrators that give different calibration levels. For example, after the instrument has been calibrated to 114 dB, it may be checked with a calibrator giving 90 dB. The results should be within ± 0.5 dB.

3 Testing time constants The time for the displayed sound pressure level to decay depends on whether the fast, slow or impulse time weighting is selected, but should always decay at the rate of 4.3 dB/time constant. To test that the time weightings are working correctly the time should be measured for the display to drop a fixed number of decibels immediately after the calibrator is switched off. The results may be substituted into the following formula to verify that the instrument is working correctly:

$$\text{Time constant} = \text{time to decay } n \text{ decibels} \times 4.3/n \text{ seconds}$$

The time constants should be:

1   Fast weighting: 0.125 seconds.
2   Slow weighting: 1.0 seconds.
3   Impulse weighting: decay between 1.5 and 3.0 seconds.

4 Testing 'A' weighting A pistonphone usually produces a fixed sound level at 250 Hz. The 'A' weighted level is 8.6 dB less than the linear level at this frequency, and so a change from linear to 'A' weighting should result in a decrease of 8.6 dB on the display. At 1 kHz, there is no difference between the 'A' weighted and linear levels, and so when an electroacoustic calibrator is attached no difference should be registered on the display when changing from linear to 'A' weighting.

5 Testing L_{eq} or L_{Aeq} Doubling the exposure time for a constant noise level increases the L_{eq} (or L_{Aeq}) by 3 dB and, similarly, halving the exposure time for a constant noise level decreases the L_{eq} by 3 dB. The L_{eq} function of the instrument may therefore easily be tested by switching the calibrator on for fifteen seconds and then off for fifteen seconds, after which the displayed level should drop by 3 dB. The decay of the displayed L_{eq} against time should be as follows:

Time (seconds)		Displayed L_{eq}
0	Calibrator on	Calibration level
15	Calibrator off	Calibration level
30	Calibrator off	Calibration level − 3 dB
60	Calibrator off	Calibration level − 6 dB
120	Calibrator off	Calibration level − 9 dB
240	Calibrator off	Calibration level − 12 dB

6   *Testing L_{AE} (SEL)*   The L_{AE} is equivalent to a one-second L_{Aeq} and may be calculated from the following equation:

$$L_{AE} = L_{Aeq} + 10 \lg T/T_0$$

where T is the time taken for the measurement (seconds), and T_0 = one second.

To check that the L_{AE} function is working correctly the calibration level should be measured for ten seconds. The display should show the calibration level + 10 dB, since $10 \lg 10 = 10$.

7   *Testing L_n and L_{An}*   The L_{10} is the sound pressure level which has been exceeded for 10 per cent of the measurement period, and may be checked by switching the calibrator on for 10 per cent of the nominal measuring time, after which the display should indicate the calibration level. For example:

Total measuring time 300 seconds
Calibrator on for 10 per cent of time = 30 seconds
Calibrator off for 90 per cent of time = 270 seconds
L_{10} display (after 30 seconds) = calibration level.

Similarly, the L_{50} may be checked by turning the calibrator on for 50 per cent of the nominal measuring time, L_{90} by turning the calibrator on for 90 per cent of the time, and so on.

Factory calibration

Both the instrument and calibrator should be returned regularly to the manufacturer or an accredited calibration laboratory for testing and calibration. The time interval should be in accordance with the relevant standards, but should be at least every five years for instruments and every year for calibrators. Where an instrument is being used for cases of litigation, it should be factory calibrated at least every two years and preferably every year.

Sound power measurements

Sound power levels can be calculated from sound pressure or sound intensity levels in several ways depending on the measurement environment. See ISO 3740 to 3746[2] inclusive.

Free field

Free field measurements can be carried out where reflective surfaces are not present to influence results, for example on open ground with no nearby buildings and very low background noise, or in an anechoic chamber. Sound pressure levels are recorded at certain prescribed points around the sound source (ISO 3745)[3] at a set distance from the source. At least six measurement positions should be taken, preferably more. The sound pressure levels measured are logarithmically averaged, and the sound power can then be calculated from

$$L_w = L_p + 20 \lg r + 11 \qquad \text{dB}$$

(for spherical sound radiation) or

$$L_w = L_p + 20 \lg r + 8 \qquad \text{dB}$$

(for hemispherical sound radiation);

where L_w = sound power level, dB re 1 pW
        L_p = average sound pressure level, dB re 20 μPa
        r = distance between source and measurement positions, m.

Reverberant room

If the sound source is placed in a highly reverberant room, provided the measurement positions are not too close to the source or a wall, the sound power level (ISO 3741/3742)[4], L_w, can be estimated from the formula

$$L_w = L_p + 10 \lg V - 10 \lg T - 14 \qquad \text{dB}$$

where V = volume of room, m³
        T = reverberation time of room, s
        L_p = average sound pressure level, dB re 20 μPa.

Sound intensity measurement

Acoustic intensity analysis is a powerful practical tool for noise source detection, sound power measurement and two- and three-dimensional sound field mapping. Sound pressure level measurements yield only amplitude information and are made using a single microphone; sound intensity measurement is achieved by the use of two near-perfectly matched microphones mounted face to face in a special probe with a constant gap between them. This method of measurement allows the direct calculation of both the magnitude and direction of the sound intensity field at a point in space.

This technique may be used, for instance, to identify the source of a particular tonal component where an item of equipment produces noise by a variety of mechanisms, even though there may be considerable masking noise present. This is usually achieved by holding the probe close to the surface of the machine and moving it over the surface whilst noting the magnitude and direction of the intensity field on the display of the analyser. The signal to be investigated should be reasonably steady, but if this is not the case, signal averaging may be performed over a suitable frequency range for an appropriate time period.

Sound power levels of a variety of noise sources may be easily obtained using the intensity technique since no special test room is required, and the source may be investigated *in situ*. A notional surface is established which completely surrounds the object under test, and Gauss's theorem predicts that the total intensity measured over the plane surface will be due only to the source contained within it. The effect of extraneous signals and random noise tend to cancel out, so large turbines, generators and the like may be investigated on site without recourse to a special acoustic environment such as an anechoic chamber or reverberant room. Figure 2.4 shows a model of a sound intensity analyser.

Instrumentation

Normally an acoustic measuring instrument will consist of a detector (the transducer), voltage amplifiers, signal processors and an output display, which may be of an analogue or digital form.

The detector

The function of this device is to convert the input quantity, which is usually acoustic pressure, into an electrical quantity, typically voltage, for easy signal processing. Different detectors are available depending on the application, for example strain gauges are used for stress measurements and hydrophones for underwater acoustics. In this section the two detectors which will be discussed are the capacitor (condenser) microphone and the accelerometer.

The ideal acoustic detector should operate as a linear transducer, that is, the output should be directly proportional to the input. The main requirements for such a device are wide dynamic and frequency ranges (200 dB and dc to 100 kHz respectively), good linearity and a high signal to noise ratio.

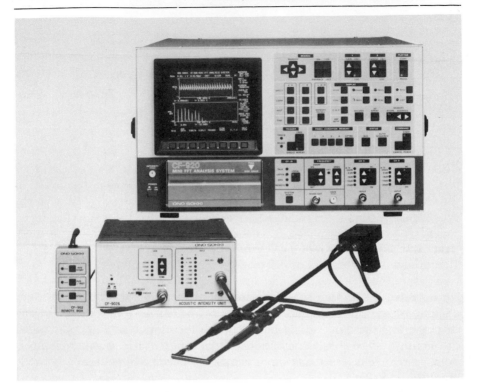

Figure 2.4   *Sound intensity analyser*
(Reproduced by permission of Lucas CEL Instrument's Ltd.)

For precision acoustical pressure measurements the capacitor (condenser) microphone is preferred because of its good linear frequency and dynamic response. There are two types of capacitor microphone which are available; the air capacitor microphone and the electret microphone.

The air capacitor microphone

Figure 2.5 shows a sectional view of the microphone cartridge. The microphone consists essentially of a thin metal foil diaphragm separated by a small distance from a parallel backplate. The diaphragm and backplate are electrically insulated and form the plates of a capacitor. A dc polarizing voltage (commonly 200 V) is applied across the plates so that the microphone operates as a capacitor. Disturbances in the acoustic pressure will cause minute displacement of the diaphragm from its rest position. This variation in the separation between the plates will cause a change in the capacitance, which results in a temporary shift in the polarizing voltage. This time-varying voltage from the microphone is directly proportional to the input sound pressure. The air

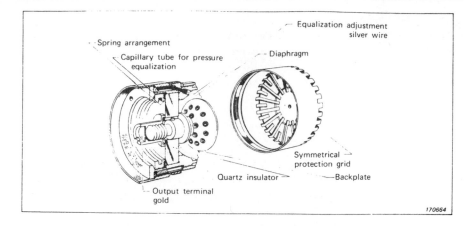

Figure 2.5   *Sectional view of a capacitor microphone cartridge*
(Reproduced by permission of Bruel and Kjaer (UK) Ltd.)

gap is vented to the atmosphere to prevent changes in the ambient pressure affecting the static position of the diaphragm. The efficiency of the microphone for converting the input sound pressure into the output ac voltage is called the *sensitivity*. The formal definition of the sensitivity of a microphone is the output ac voltage per unit input acoustic pressure and is usually between 30 mV/Pa and 50 mV/Pa.

The sensitivity of the microphone depends on the dc polarizing voltage, the surface area of the diaphragm and the separation distance between the plates. These factors are important when choosing a microphone for a given task. A microphone of 12.5 mm (0.5 inch) diameter is suitable for most applications because of its wide dynamic range, typically sound pressure levels from about 20 dB to about 150 dB.

When dealing with low sound pressure levels, high sensitivity is required. The 25mm (1 inch) microphone is the best option for this particular case. Conversely, if high sound pressure levels are investigated (say above 160 dB) a smaller diameter microphone is preferable, say 6mm. Such a microphone may also be very effective for high frequency investigations.

The design and materials selected for the diaphragm and the microphone housing are chosen to provide good temperature stability in the range − 10°C to 50 °C.

The diaphragm should never be touched and should be cleaned only with extreme care. If the air capacitor microphone is subjected to extreme changes in temperature (for example left in a car overnight in winter and then taken into a warm room) condensation may occur between the diaphragm and the backing plate. Switching on the dc polarizing voltage at such time can result in a damaged diaphragm. This may be avoided by warming the microphone before use. If the microphone should become wet, it can be dried by the application of gentle heat.

The electret microphone, which is becoming more widely used, is similar in construction to the air capacitor microphone, but a prepolarized polymer is bonded to the backplate and with the air gap forms the dielectric of the capacitor. The result is a microphone which is better protected against humidity but to achieve IEC 651[5] Type 1 accuracy will be more expensive than the air capacitor type.

The piezoelectric accelerometer

The most common transducer used for vibrational measurements is the piezoelectric accelerometer, because of its wide dynamic and frequency range combined with a good linear response. Figure 2.6 provides a schematic diagram of one.

The construction of such a device invariably consists of a slice of piezoelectric material, such as an artificially polarized ferroelectric ceramic. When such a material is mechanically stressed under tension, compression or shear, a small electric charge is generated at its pole faces which is proportional to the applied force. Normally the piezoelectric material is mounted as a sandwich between a relatively massive inertial base and a spring loaded mass. When the assembly is vibrated the mass applies a force to the piezoelectric element which is directly proportional to the vibratory acceleration (the input).

There are two basic types of accelerometer available, the shear accelerometer and the compression accelerometer, the difference between the two being the direction in which the force is applied to the piezoelectric element. The shear type accelerometer has good universal characteristics and so is the more frequently used, but the compression type may be employed when more specialist measurements are to be made, where such characteristics as low mass, high sensitivity, high temperature and shock capability are important.

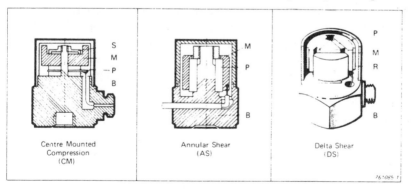

Schematic of B & K accelerometer configurations
S = Spring   M = Mass   B = Case   C = Cable
P = Piezoelectric element   R = Clamping Ring

Figure 2.6   *Schematic accelerometer configurations*
(Reproduced by permission of Bruel and Kjaer (UK) Ltd.)

The sensitivity of an accelerometer can be expressed as the charge sensitivity, which for convenience is normally quoted in terms of picocoulombs per acceleration due to gravity (for example 98.1 pC/g). Alternatively, the voltage sensitivity may be given (for example 72.9 mV/g). The sensitivity will depend on the piezoelectric material and the size of the mass. To obtain a high sensitivity a large mass should be used but this reduces the operating frequency range. The useful frequency range of a transducer may be taken as one-third of its resonant frequency; for example, if the resonant frequency is 30 kHz, measurements of up to 10 kHz may be made, with an accuracy of approximately ± 1 dB.

The accelerometer should be firmly mounted on the vibrating surface to ensure that there is maximum transmission of vibration energy. There are several ways in which this can be achieved depending on the nature of the vibrating surface. It must be noted that the dynamic and frequency ranges of the mountings must be known for the acceleration to be accurately measured. Careless mounting may result in the reduction of the resonant frequency, thus decreasing the frequency range, and may possibly cause damage to the accelerometer.

Some of the ways in which the accelerometer can be mounted to a surface are by use of a magnet, by cementing with an adhesive or a thin layer of wax, or by the use of a threaded stud. For non-critical purposes a hand-held probe may be employed. The most convenient of these is mounting by a magnet, but for obvious reasons this is not always possible. Mounting with a threaded stud, which can be electrically isolated to prevent earth loops interfering with the accelerometer, is the most reliable and gives the best results, but requires the vibrating surface to be drilled to accommodate the stud.

Accelerometers are generally robust and can be usefully employed in a number of inhospitable environments. For example, they are usually tightly sealed and so, provided the connecting cables are also properly sealed, may be used in humid environments or even within liquids for short periods. Many accelerometers will withstand high temperatures (up to 250°C or higher) and so are used extensively for plant and machinery health monitoring. Unfortunately, fluctuations in temperature can affect the accelerometer's output, so care must be taken when selecting the accelerometer appropriate for the task, for example the shear accelerometer has small sensitivity to temperature transients. There are many other environmental conditions which could affect the operation of the accelerometer, such as electromagnetic noise and nuclear radiation, so it is very important to be aware of any possible sources of interference before starting a programme of measurements.

The sound level meter

For simplicity, the principle of the sound level meter is described as for the measurement of airborne sound only. The microphone detects any fluctuating changes in the ambient pressure and converts these into an electrical quantity.

The voltage signal produced is proportional to the measured acoustic pressure provided the microphone is operating within its linear regime. The detected acoustic signal can be in many forms; three common types are pure tones, broad band noise and impulsive noise. The type of noise will have some effect on the signal processing and analysis that is to follow.

The voltage signal from the microphone must be amplified before any signal processing can be performed. After this initial amplification the average intensity or rms sound pressure of the acoustic radiation is determined by analogue electronic averaging circuits. The averaging times of these circuits are selected according to the type of noise to be measured and the relevant standards. Normally at least two, but sometimes as many as four averaging times are available on the meter: see Table 2.2 for standardized time constants.

The next stage in the signal processing is to quantify the voltage equivalent of the rms sound pressure level. This is achieved by means of a second amplifier whose output is the common logarithm of the input, and the resulting voltage is displayed on a moving coil meter or a digital display calibrated in terms of sound pressure level.

It is important to remember that the rms level is an averaged result which may contain components many times larger than its own value. This means the maximum pressure peaks incident on the transducer could be overloading the meter and so exceeding the full scale deflection (FSD) even though the rms value is mid-scale.

The *crest factor* is a ratio which indicates how large the measured peaks are compared to the rms level. For a pure tone the ratio of peak to rms is 1.414:1 or 3 dB, but for wide-band random noise the crest factor can be between 3:1 and 10:1 (that is, 10 dB to 20 dB). For impulsive noise the crest factor can be as high as 1000:1 (60 dB). This illustrates the importance of the various sound level meters in regard to the performance of the rms circuits. Crest factor requirements of the current equipment standards are shown in Table 2.3.

When a sound level meter is overloaded it is usual for errors to arise at the electronic averaging circuit generating the rms signal. Overload detectors are placed on the input to indicate this to the operator although, strictly,

Table 2.2

Meter averaging times

Function	Time constants
Slow	1 s
Fast	125 ms
Peak	100 μs
Impulse	35 ms rise/1.5 s decay

Table 2.3
Crest factor requirements of current equipment standards

Sound level meter	Crest factor	Standard
Type 2	10 dB	IEC 651
Type 1	15 dB	IEC 651
Type 1 I	20 dB	IEC 651
Type 1 (impulse capability)	63 dB	IEC 804

the relevant standards require such detectors only on integrating and impulsive instruments. It can be seen that an instrument with an inadequate crest factor will not be able to include all the signal in the calculation of the rms level and hence an incorrect answer will be displayed. With signals containing many peaks, overloads can occur when the average level is still well on scale. Figure 2.7 shows a schematic diagram of a sound level meter.

If the visual output is a moving coil galvanometer the FSD will tend to be limited to 20 dB (that is, a range -10 dB to 10 dB). With such meters an attenuator set in 10 dB increments precedes the galvanometer and the working range of the instrument may thus be 20 dB to 140 dB, the attenuator being set from 0 to 130 dB. Thus, if the meter is overloading on the 70 dB range the solution may simply be to switch the attenuator to 80 dB.

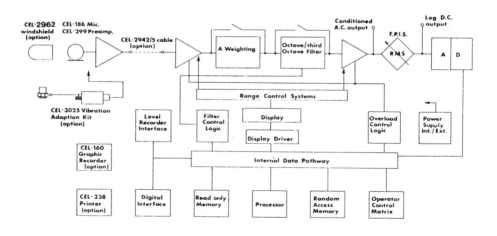

Figure 2.7 *Schematic diagram of sound level meter*
(Reproduced by permission of Lucas CEL Instruments Ltd.)

Sound level meters with digital displays often have a sound level range from 50 dB to 150 dB without incremental steps. This means that there need be no concern about overloading unless the peaks in the sound pressure level approach 150 dB re 20 μPa.

Sound level meters are now sufficiently developed that the main limitation on performance is determined by how rapidly the meter can respond to short-duration high-amplitude transients. The IEC 804[6] standard requires the system to respond to 63 dB, 1 ms pulses. It is possible for some sound level meters to process correctly a 98 dB, 1 ms pulse.

Frequency weighting networks that are required for psycho-acoustic or other reasons must be included in the ac signal path and so must be placed prior to the rms averaging circuit. Since the dynamic range of most filter networks in current use is limited, they cannot be located at the front of the measurement chain but are usually situated between the input amplifier and the main amplifier. Range changing using attenuators may therefore occur either side of the filter network. The use of highly selective networks such as the standardized 'A' weighting and octave band filters introduces complications into the overload detection arrangements because a weighting network generally deletes a significant proportion of a signal, which might otherwise overload the second amplifier. These difficulties are partly overcome by inserting dual overload detectors that detect pre-filter and post-filter overloads. If dual overload detectors are not available, it is important to set the measurement range using the linear setting, and then change to the required filter network without changing the measurement range. This procedure will prevent undetected overloads from causing an error.

The accuracy of moving coil displays depends on the judgement of the operator. In the case of meters with digital displays, the resolution becomes a function of the analogue to digital converter and its sampling rate. The fastest practical rate at which it is possible to read the display is twice a second, hence this would indicate the maximum speed. As it is difficult to form an impression of the level variation in such a short amount of time, it is now common practice to provide an analogue bar graph in addition to the main display. These are usually updated at a rate of 125 ms (8 Hz) to give a visual impression of continuity. For statistical processing and integration, considerably faster sample rates are required which are typically 60–350 Hz (17–3 ms).

Instrument selection

The reason for making noise measurements should be clearly understood before a measuring instrument is selected or a measurement procedure initiated. Some typical situations are listed below.

1   The assessment of employees' risk of hearing damage at their workplace.
2   To verify a subjective assessment of community noise annoyance.

3   To ensure that a product is within quality assurance specifications before despatch.
4   To improve a product's design and performance.
5   To predict and prevent machine failure.
6   To aid the design of a building and building services.

To select the appropriate instrument it is necessary to establish the degree of accuracy required of the measurements. The tolerances suitable for various applications are given in Table 2.4.

Full specifications of sound level meter tolerances may be found in IEC 651.[5] For frequencies outside the range given in Table 2.4, and for other limitations such as change in sensitivity with direction, IEC 651 should be consulted.

Recorders

Environmental noise analysers

Where long-term noise monitoring is required (days, weeks or months) a hand-held sound level meter is not practical and an environmental noise analyser would be a suitable instrument. Such an instrument is designed to be left out of doors in all weathers and normally includes a built-in printer to ensure that data are not lost even in the case of power failure.

Environmental noise analysers will normally record, simultaneously, many of the usual noise descriptors, such as L_{eq}, L_{AE} and several L_n values. The printout may be set for time periods ranging from one second to one hour or more. These analysers are designed to have a long battery life, typically giving up to seven days' operation with a printout every hour. A threshold detection circuit is usually available, which may be used to start and stop a tape recorder remotely when preset sound levels are detected.

Table 2.4
Instrument selection tolerances

Application	IEC type	Tolerance dB 125 Hz to 1 kHz
Laboratory use only	'0'	+ 0.7
Where litigation is possible	'1'	+ 1.0
For internal assessment only	'2'	+ 1.5
As an approximate check	'3'	+ 2.0
For educational use only	unspecified	unspecified

Graphic level recorders

When a permanent, continuous record of sound pressure levels against time is required a graphic level recorder may be used. This may be connected to the output of the measuring instrument and will produce a graph of the measured sound level on a continuous paper strip. Some types of recorder have a built-in sound level meter to which a microphone and preamplifier may be directly connected. Filter sets are also available, or the recorder may be used in conjunction with a frequency analyser, to produce a printout of sound level versus frequency in octave or one-third octave bands.

Older types of level recorder, still in current use, produce the trace by means of an inked pen on paper, but in the newer types this is replaced by an electro-discharge combination giving a faster response time. Some employ the expansion mode of operation for continuous recording, whereby the paper speed is set on slow but a faster speed is triggered by a preset sound level. Figure 2.8 provides an illustration of a graphic level recorder.

Tape recorders

The object of introducing a tape recorder into the measurement chain is to

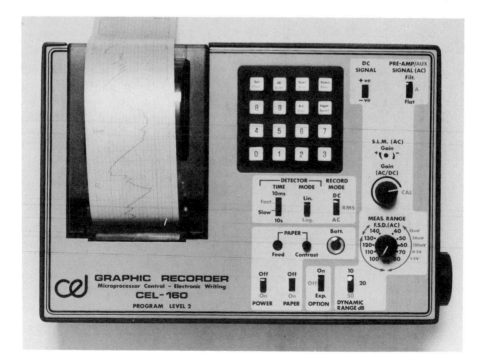

Figure 2.8 *Graphic level recorder*
(Reproduced by permission of Lucas CEL Instruments Ltd.)

capture the signal for later analysis. This could be either time history or frequency analysis or the recording may be simply for source identification. Each of these requirements places different constraints on the recording system but the important parameter is that there is an accurate reconstruction of the signal, that is, the relation between high and low frequency and level.

There are two main types of tape recording, analogue and digital. At present the former is almost exclusively used for instrumentation applications but it is expected to be soon replaced by digital recording. In analogue tape recording, time is represented by the tape travel past the record/replay head, intensity by the degree of magnetization and sign by the polarity of magnetization. Unfortunately error compensation is individually tailored for each tape recorder so a limit on the accuracy of measurements is imposed if the noise is recorded on one machine and played back on another. There is always a compromise between frequency response and tape play times, hence running speed must be selected for the application in question.

Dynamic range is determined by the maximum magnetization that may be permanently induced on to the tape (magnetic saturation) and the random noise produced by the individual particles (random magnetism). These two parameters are linked by the design compromises and any action to improve one tends to result in a deterioration of the other, for example making the particles larger to allow a higher magnetization results in an increased noise floor. There is therefore a fixed dynamic range of around 55 to 60 dB for this type of recording.

Digital tape recording relies upon the high-speed conversion of the time series signal into a sequence of discrete digital numbers. The amplitude of the signal is determined by the magnitude of the number and the polarity by its sign. Dynamic range is a function of the word length and upper frequency is set by the sample speed. If very low distortion is required high-speed sampling systems must be included in the recording system. At present digital systems are expensive but have a very wide dynamic range.

A tape is calibrated by recording a known sound pressure level and noting the level and its position within the tape's dynamic range. It is good practice to record the calibration tone at the beginning and at the end of the recording. For critical applications it is possible to record the reference signal along the complete length of the tape on the spare track of a stereo machine (it is also possible to cascade the two tracks of a stereo system to improve the dynamic range). It is of course necessary to make a note of any changes in the setting of the sound level meter attenuator switch between the recording of the calibration tone and the actual recordings. For example, increasing the sound level amplifier gain by increasing the input attenuator setting (which reduces sound level measurement range) will boost the signal level and make the recording seem louder relative to the calibration tone. Automatic level-control systems and dynamic range enhancements systems (Dolby) should not be used for instrumentation recordings and the calibrated step attenuators should be used only for gain changing after calibration. Crest factor allowance should be borne in mind when using sound level meters, since the recording level meter on the tape recorder will indicate only peak levels or sine wave rms. When the source instrument has a wider dynamic range than the recorder it is necessary to decide which part of its output is to be recorded. On playback, the calibration tone should be set by means of the output level and range switch such that it produces an output on the analysing instrument the same number of decibels below FSD as the calibration tone was below tape saturation. The tape recorder cannot then drive the analysing instrument into overload.

Outputs

Most signal processors have an inbuilt display in the form of a moving coil meter, liquid crystal display (LCD) or a cathode ray tube (CRT). In addition, many instruments are provided with a number of outputs for connection to auxiliary devices such as tape recorders, computers and printers. These may be any of the following.

1 *Unconditioned ac output.* This is usually the transducer signal without range or frequency correction. Sensitivity is generally quoted in mV for full scale deflection.
2 *Conditioned ac output.* A signal taken and affected by all the range changing functions and frequency weightings. Sensitivity is again quoted in mV for full scale deflection.
3 *Dc outputs.* The rms signal which may be linear in volts or a logarithmic form of the voltage. Linear signals are in mV for FSD and logarithmic signals are in mV/dB.
4 *Filter connections.* These provide an input/output facility for the connection of external filters.
5 *Digital outputs.* These may be for computer and/or printer connection. They may be serial (RS-232C) or parallel (IEEE-488) conforming to standardized protocols, or specially constructed to suit a manufacturer's particular design requirement.

References

1 ISO 266 (1975) *Preferred Frequencies for Acoustical Measurements*, ISO.
2 ISO 3740 (1980) *Acoustics – Determination of the Sound Power Level of Noise Sources – Guidelines for the Use of Basic Standards and for the Preparation of Noise Test Codes*, ISO.
3 ISO 3745 (1979) *Acoustics – Determination of the Sound Power Levels of Noise Sources – Precision Methods for Anechoic and Semi-Anechoic Rooms*, ISO.

4   ISO 3741 (1975) *Acoustics – Determination of the Sound Power Levels
    of Noise Sources – Precision Methods for Broad-Band Sources in Reverbera-
    tion Rooms*, ISO.
    ISO 3742 (1975), *Acoustics – Determination of the Sound Power Levels
    of Noise Sources – Precision Methods for Discrete-Frequency and Narrow-
    Band Sources in Reverberation Rooms*, ISO.
5   IEC 651 (1979) *Specification for Sound Level Meters*, IEC.
6   IEC 804 (1985), *Specification for Integrating-Averaging Sound Level Meters*,
    IEC.

Chapter 3

ENVIRONMENTAL NOISE AND VIBRATION

P. T. Freeborn, and **S. W. Turner,** London Scientific Services

A wide range of noise and vibration sources may be termed environmental and, because people react differently to each one, quantifying and controlling the impact of these sources must be treated separately. In general, environmental noise and vibration covers the effect of external sources on people within buildings. Hence, in this chapter, such problems as aircraft noise and residential areas, traffic noise and schools are considered. Also covered is the noise from railways, helicopters, industrial premises and entertainments such as pop concerts.

The chapter begins, however, with the question of noise rating – the method of quantifying the problem and the means by which acousticians describe objectively what is, in fact, a subjective phenomenon.

Noise rating

Why have a rating?

Noise in the environment affects all people and, moreover, affects them sufficiently that most express an opinion about it. At one extreme, the noise can be loud enough to feel physically uncomfortable (for example in a very noisy disco) and, if persistent enough, can lead to some deterioration in health (noise-induced hearing loss). At the other extreme, the noise may be just perceptible but yet intensely annoying (the throb of a distant generator). In between, conversation can be disrupted (during an aircraft flyover), and at night, if there is too much noise, sleeping is difficult. Noise must then be considered, whether in determining the design and layout of new housing or deciding where a new by-pass should be routed and this means that the designers need the effects of noise on people to be rated and quantified.

Quantifying the noise

When approaching noise measurement, the various features of the noise that are likely to affect the subjective reaction must be considered. These include:

1   The type of noise: for example, is it continuous at a constant level, or continuous but fluctuating in level; or is it intermittent?
2   The frequency content of the noise: is it broad band; or has it a prominent frequency (that is, a tonal quality)?
3   The time of day it occurs.

The relevant combination of these features will lead to a parameter and unit to use and from that the noise exposure can be quantified.

The details of noise measurement are discussed above, but knowing a number that defines the level of exposure is not enough – the effect that exposure has on people must also be defined: Will it disturb concentration? Will there be annoyance and, if so, how much? Thus the noise exposure must be linked with a measure of the corresponding reaction of people to that noise.

Subjective reaction

Unfortunately, it is not possible to measure annoyance or disturbance directly. Therefore, elaborate questionnaires are used to try to quantify the subjective response, and generally it is the average response of a group of people to various noise exposures that is found.

Thus, in the case of aircraft noise at London's Heathrow Airport, the graph shown in Figure 3.1 was produced.

Generally, it is found that the correlation between the noise exposure parameter and annoyance is not that good when the response is described in terms of the average reaction of a group of people. This is the result of the many non-acoustical factors that will affect individuals' reaction to noise. For example, there is their attitude towards the source (for example, they may just dislike helicopters so any helicopter noise is likely to be considered awful) or their feeling about their neighbourhood or, simply, their general state of health when the judgement is made – all these can influence the response in a way that has nothing to do with the noise exposure. Consequently, many parameters and units have been proposed to describe the variety of noise that is encountered, to improve the accuracy of predicting the corresponding response. Some of these are listed in Table 3.1.

It is generally acknowledged that those parameters in current usage are not ideal as predictors, but are probably the best available. Some improvement

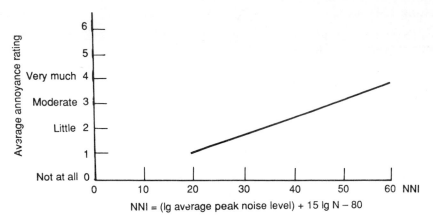

Figure 3.1   *Average annoyance vs. Noise exposure from aircraft*

Table 3.1

Some units and parameters used to quantify noise exposure

Units	
Sound pressure level (SPL)	dB or dB(Lin)
'A' weighted SPL	dB(A)
'B' weighted SPL	dB(B)
'C' weighted SPL	dB(C)
'D' weighted SPL	dB(D)
Perceived noise level	PNdB
Parameters	
Statistical percentiles	L_1, L_{10}, L_{50}, L_{90}, etc.
Equivalent continuous level	L_{eq}
Single event level	L_{AX} or SEL
Noise and number index	NNI
Noise rating/Noise criteria	NR/NC
Weighted standardized level difference	$D_{nT,w}$
Corrected noise level	CNL

in correlation has been found by looking not at the average response of the group but at the percentage seriously annoyed or disturbed at a particular noise exposure. The non-acoustical factors start to have less influence at that level of annoyance so the response is more closely related to the noise exposure.

Setting a standard

The final stage needed in noise rating is the determination of the level at which some action is required, that is, the setting of the standard or criterion. Clearly, the ideal would be to reduce the level of annoyance to zero, but economic and political constraints operate to keep this aim an ideal. A compromise must then be reached and decisions must be made regarding, for example, the noise exposure above which grants should be given for improved sound insulation near an airport or how many helicopters should be permitted to use a proposed new heliport.

Some standards have been set so that half the people affected could be expected to be satisfied. This, of course, means that even if the standard is met, half could be expected to be annoyed. Other standards are likely initially to have been set intuitively and then, over a period of time, been proved to be reasonable, and some, especially when financial grants are involved, are possibly more closely related to the sum of money available rather than level of satisfaction that could be expected. Regardless of the derivation of the standard it is essential to have one because designers and planners need to have a quantifiable basis upon which to make their decisions, even though that basis may have been developed in a not particularly precise way.

Having set the standard, there is then the question of the individual whose noise exposure falls just short of the standard, and hence, for example, receives no grant, whereas the neighbour nearby qualifies. To these people the noise will most certainly seem no different and yet the scientists say that it is. Unless the ideal of satisfying everyone is achieved, this problem of having to exclude some people will always occur.

Individual response

Although the average reaction of a group to a particular noise exposure can be predicted, the response of an individual cannot. Once, at an inquiry into a proposed airport expansion, the inspector asked the question: 'There are five people in the high street of the nearby town. One is seriously annoyed by aircraft noise, three are moderately annoyed and one is not annoyed. If the expansion proceeds how will their reaction change?' Even with the information about the existing and predicted future noise exposure and data such as in Figure 3.1, this question could not be answered. It is not possible to predict the reaction of an individual. All that could be estimated was the change in the percentage of the population of the town likely to be seriously annoyed.

There are many cases of the acoustical standard being met and yet the individual involved still complains about the noise and expresses dissatisfaction. The circumstances surrounding such a complaint would need careful examination to determine, for example, whether the noise is the genuine source of the dissatisfaction or whether it is being used to express some other grievance. At all times though, the basis of the standard should be remembered for within it there is likely to be the implicit expectation that some will be dissatisfied even when the standard is met.

Road traffic noise

The Wilson Committee, in its report published in 1963,[1] concluded that 'in London (and no doubt this applies to other large towns as well) road traffic is, at the present time, the predominant source of annoyance, and no other single noise is of comparable importance'. Since then regulations controlling the noise emissions from vehicles[2] have been introduced and subsequently tightened, and a statutory procedure to reduce the effect of noise from new or improved highways[3,4] has been implemented. In that period, also, the number of vehicles using the roads in the United Kingdom has risen and there has been a continuing upward trend in complaints made about road traffic noise.[5] Clearly then, this noise source is still the most prevalent and is continuing to cause annoyance and disturbance.

It is arguably the effect of traffic noise on sleep that is the most significant, and much research has been carried out to quantify the nature and extent of sleep disturbance.[6] If there is a sleep loss, not only is this in itself likely to cause annoyance to the individual, there is also the effect of the subsequent impairment of performance as a result of the sleep loss. This can have a wide-ranging socio-economic impact with, for example, people performing less well at work as a result of exposure to high levels of noise. With the trend in vehicle usage increasing, the scale of the problem is unlikely to diminish in the near future, so the need to control and minimize the effects of road traffic noise will remain.

Character of road traffic noise

As traffic noise describes not only the occasional vehicle passing along a suburban street but also the continuous flow of a six-lane dual-carriageway motorway, it is clear that the character of the noise is not constant, yet both extremes comprise the same basic sources. An individual vehicle travelling along a road generates noise from its engine, its transmission and exhaust, the interaction between the tyres and the road surface and the interaction between the vehicle and the air it is passing through. All these sources can effectively be combined and considered as a moving point source. The propagation law covering this type of source means that a 6 dB(A) reduction for every doubling of distance away (6 dB/dd) should be expected.

The level of noise generated by this vehicle will depend not only on its size, type and shape but also on the speed it is travelling, whether or not it is accelerating, and whether the road is on a gradient. When there are many vehicles using a road at the same time, the collection of point sources starts to merge and effectively behaves as a line source. This means that the reduction becomes 3 dB/dd and for freely flowing traffic this line source assumption holds well.

Figure 3.2 shows the variation in noise level with time at two locations. Site A overlooked a three-lane urban motorway, whereas site B was close to a relatively quiet suburban road. It can be seen that although the general level is fairly similar at both sites the character is different. The level is similar because the monitoring position was much closer to the suburban road than to the motorway. At site A the noise level is fairly constant and remains within a range of about 10 dB(A), and it is not possible to identify an individual vehicle. At site B the peak levels from individual vehicles are more distinct (even a car horn) and the range of noise level is over 20 dB(A).

Figure 3.3 shows the frequency spectra for each site and here it can be seen that the two sites are fairly similar. Even with a wet road, although there are changes in the frequency spectrum, the overall dB(A) value has been found to remain constant.[7] As a result, it has been possible to develop

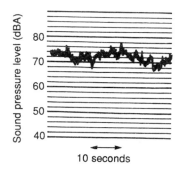

Site A: Overlooking a 3-lane urban motorway

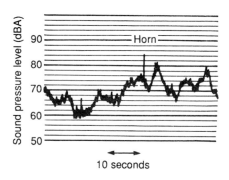

Site B: Fairly close to a 'quiet' suburban road

Figure 3.2   *Variation in noise level with time at two different sites*

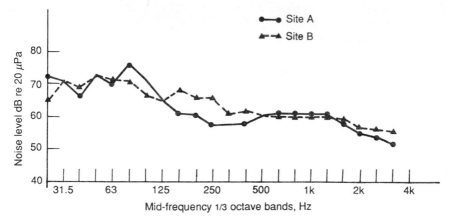

Figure 3.3   *Frequency spectrum of the traffic noise shown in Figure 3.2*

relatively simple prediction techniques for road traffic in terms of the dB(A) unit with a fairly good degree of reliability.

Measurement of traffic noise

The two graphs in Figure 3.2 highlight the problem of finding a suitable descriptor for traffic noise. With a level that can be either fairly constant or greatly varying, a statistical parameter is used. By sampling the noise level, the cumulative distribution can be found and from that the level exceeded for 1 per cent, 10 per cent, 50 per cent or 90 per cent (or any other percentile) of the measurement period calculated. These percentiles are described as L_1, L_{10}, L_{50} and L_{90}, and so on. For the sites represented in Figure 3.2, the distribution in noise level is shown in Figure 3.4, and the corresponding statistical parameters given in Table 3.2. Also shown in that table is the value of the equivalent continuous noise level (L_{Aeq}).

It has been found that it is the L_{A10} that correlates most closely to the subjective response with freely flowing traffic. So, at these two monitoring positions, it would be expected that the average reaction of a group of people to the traffic noise would be similar, as the two levels are almost the same.

The different character of the noise at the two sites is quantified by the L_{A90} values, where there is a 7 dB(A) difference. At site A, the relatively constant nature of the noise level is shown by there being only a 4.6 dB(A) difference between the L_{A10} and L_{A90}. At site B this difference is 13 dB(A), Although there are longer periods of relative quiet. This variability in the noise level which probably results in there being a similar level of annoyance. The Traffic Noise Index was developed to reflect this variability and to relate the overall noise level to the background level.[8] It has, however, a fairly limited range of application, and it is the L_{A10} that has been used in the

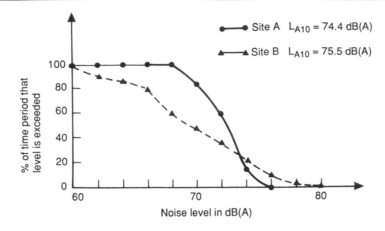

Figure 3.4   *Cumulative distribution of traffic noise shown in Figure 3.2*

Table 3.2
Values of statistical parameters for traffic noise shown in Figure 3.2

Percentile	Levels in dB(A)	
	Site A	Site B
L_1	76.3	80.0
L_{10}	74.4	75.5
L_{50}	72.5	70.0
L_{90}	69.8	62.5
L_{Aeq}	72.4	72.1

legislation concerning traffic and, more particularly, the value of L_{A10} averaged hourly over the time period 0600–2400 hours ($L_{A10}(18)$).

Figure 3.5 shows the relationship between the subjective response and the $L_{A10}(18)$ level.

With the advance of the capability of acoustic equipment there are now instruments that will measure L_{A10} directly. Measurements can be made either under free field conditions (Figure 3.6(a)) or at a facade (Figure 3.6(b)). It is found that the facade measurement yields levels approximately 2.5 dB(A) higher than the equivalent free field condition, due to reflection effects.

It is also worthwhile obtaining the value of the L_{Aeq} when assessing traffic noise, as some standards are now being specified in terms of this parameter. Although L_{Aeq} can be measured directly, if only the L_{A10} is known the following approximation can be used for freely flowing traffic:

$$L_{A10} = L_{Aeq} + 3 dB(A)$$

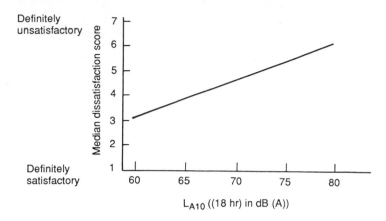

Figure 3.5 *Dissatisfaction with traffic noise vs. L_{A10} (in dB(A)) averaged hourly between 0600 and 2400 hours.*

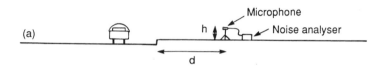

Figure 3.6(a) *Free field measurement of traffic noise*

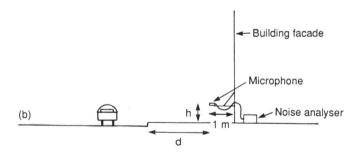

Figure 3.6(b) *Facade measurement of traffic noise*

The L_{A10} is usually measured over hourly periods and then the arithmetic average of those hourly values found for various periods depending on the requirements. Some examples are given in Table 3.3.

It has been found that there are fairly consistent relationships between some of these parameters. For example:[9]

Table 3.3
Uses of different parameters

Parameter	Period	Use
L_{A10} (18)	0600–2400	Legislation
L_{A10} (12)	0700–1900	Living room standards
L_{A10} (8d)	0900–1700	Office standards
L_{A10} (7)	0900–1600	School standards
L_{A10} (2)	2200–2400	Bedroom standards
L_{A10} (8n)	2200–0600	Sleep disturbance standards

$$L_{A10}(8d) = L_{A10}(18) + 1.5\text{dB(A)}$$

$$L_{A10}(7) = L_{A10}(18) + 1.5\text{dB(A)}$$

Furthermore, it is possible to estimate the $L_{A10}(18)$ by measuring over just part of the time:

$$L_{A10}(18) = L_{A10}(3) - 1\text{dB(A)}$$

where $L_{A10}(3)$ is any three consecutive hours between 1000 and 1700.[10]

For fairly busy roads (that is, where there is only the occasional period with no traffic) a twenty-minute measurement will be a good approximation to the corresponding hourly value.

To determine the effect of a single road on a fairly wide area, it is not necessary to measure at all the points of interest. Instead, it is possible to use a combination of measurement and prediction.

Prediction of traffic noise

When planning new roads, it is important to be able to predict the noise level that is likely to be generated so that measures can be taken to minimize the environmental impact. Extensive work has been carried out to enable the L_{A10} level at the facade of a building to be estimated from traffic flow data. The standardized procedure is set out in detail in *Calculation of Road Traffic Noise* (CRTN),[10] where a number of examples are given.

From the information about the traffic flow, a basic noise level (BNL) is found either as an hourly L_{A10} or $L_{A10}(18)$. This is effectively the noise level 10m from the kerbside. The propagation corrections are then applied with a line source being assumed, 3.5m from the kerb and 0.5m high. The effect of the ground conditions and whether the view of the road from the reception point is obstructed are included, with, finally, any corrections to account for reflections. For most purposes the noise level is calculated to the nearest 0.1 dB(A). CRTN gives charts of the corrections so that the values can be found graphically. As an alternative, equations for the various factors have been derived which can be used within certain limits. These have enabled the development of computer-based techniques, which have simplified the prediction work. The details of the site, including the location of the road, buildings, barriers and the general topography are entered into the computer, with the use of digitization, along with details of the traffic flow. The calculation program then finds the noise level at all the facades of interest. This method offers great flexibility as it enables the effects of changes in traffic flow, barrier height and road alignment to be easily found.

A revision of *Calculation of Road Traffic Noise* was published in 1988. The main principles of the techniques have remained unchanged, but some of the detailed methods have been altered regarding, for example, the propagation of sound over soft ground, the correction for the type of road surface, the effect of barriers and situations where the traffic flow is low.

Legislation and standards

Under the provisions of the Land Compensation Act 1973[3], the Noise Insulation Regulations were brought into operation that year, and subsequently revised in 1975.[4] These lay down a procedure for determining whether people living near a new or improved highway are eligible to receive compensation in the form of a package of secondary glazing and mechanical ventilation. The criteria to be satisfied for dwellings and other types of residential buildings within 300m of the nearest point of the new or improved highway are:

1 The total expected maximum traffic noise level L_{A10} (18) must not be less than 68 dB(A) at one metre from the facade.
2 There must be an expected increase of not less than 1.0 dB(A) in total traffic noise from highways in the vicinity compared with total traffic noise before the works to improve or construct the highway were begun.
3 In combining the expected maximum overall level of traffic noise from the new or altered highway and other highways in the vicinity, the new or altered highway must effectively contribute not less than 1.0 dB(A) to the increase in traffic noise.

The 'expected' increase is the likely noise level within fifteen years of the new or improved highway opening.

As the highway authority has to pay for this work, it is in its interest to make use of prevention techniques (see 'preventive measures' below) to minimize the resulting facade noise level. The highway authority is also given discretionary powers to extend eligibility along contiguous facades so that the limit of eligibility does not end between two adjacent dwellings in a terraced block. Subjectively, there would seem to be no difference in noise exposure at those two properties, so the regulations permit this flexibility.

The problems of having to set a standard discussed above under 'Noise rating', can be seen here. First, the criterion is 68 dB(A) L_{A10} (18), which is at a median dissatisfaction score (see Figure 3.5) of about four, that is, only about half the people disturbed by traffic noise would receive compensation. Second, there is the anomaly of an individual who suffers an increase in L_{A10} (18) from 77.7 to 78.6 dB(A) not receiving the package, whereas an increase from 67.7 to 68.7 dB(A) would qualify. Although the noise level in the first case is 10 dB(A) higher, no additional insulation is available. Some authorities are, however, trying to resolve this problem by using their discretionary powers.

Following the revision of *Calculation of Road Traffic Noise*, a revised set of Noise Insulation Regulations is expected to be laid before Parliament towards the end of 1988. The criteria for eligibility are expected to remain unchanged,

The first traffic noise standard was effectively set in the Wilson report,[1] which suggested that the L_{10} levels set out in Table 3.4 should not be exceeded.

In 1973 the Department of the Environment published Circular 10/73, *Planning and Noise*,[11] which recommended that:

1   Sites exposed to an L_{A10} (18) of more than 70 dB(A) should wherever possible not be used for new housing.
2   It is desirable to restrict new development to levels substantially lower than 70 dB(A).
3   Preventive measures, such as barriers, should be used in urban areas where an L_{A10} (18) of 70 dB(A) cannot be avoided.
4   It is essential that the building specification be such that in no dwelling is the internal L_{A10} with windows closed greater than 50 dB(A). This is a minimum standard – an internal level of 40 dB(A) can be regarded as a 'good' standard.

This circular specifies the units with the windows closed, but unlike the Noise Insulation Regulations, makes no mention of the need to provide an alternative means of ventilation.

The Greater London Council's (GLC) *Guidelines for Environmental Noise and Vibration*, adopted in 1976[12] state that for living rooms in new dwellings, the L_{A10}(12) should not exceed 50 dB(A); and in bedrooms the L_{A10} (2) should

Table 3.4

Recommended maximum L_{10} levels (Wilson)

Location	Living room (0800–1800)	Bedroom (0100–0600)
Country area	40 dB(A)	30 dB(A)
Suburban (away from main traffic routes)	45 dB(A)	35 dB(A)
Busy urban	50 dB(A)	35 dB(A)

not exceed 35 dB(A). The period of 2200–2400 hours was chosen as it was found that this was the time when most people went to bed, and one of the problems caused by high levels of traffic noise is difficulty in falling asleep.

The bedroom criterion is very difficult to achieve in many urban situations if the bedroom overlooks the road, and it is essential to make use of the techniques of layout (described in 'Preventive Measures', below) to overcome the problem. Consequently, for existing dwellings a target value of 40 dB(A) is used.

These guidelines are in the process of being revised and are likely to make more use of the L_{Aeq} parameter. In addition, for bedrooms L_{A1} may be used, and the time period for assessment may become the worst hour in the evening (1900–2200 hours) and the night (2200–0700 hours).

A 1987 standard[13] recommends the following maxima for steady intrusive noises:

1   bedrooms – $L_{Aeq} = 30 - 40 dB(A)$;
2   living areas (for conversation and listening to the radio and television) – $L_{Aeq} = 40 - 45 dB(A)$.

The time period is not specified, although it is implied that the worst appropriate period should be used (for example one hour). More detailed work[6] has found that sleep disturbance corresponds well with the L_{A10} (2200–0600), and that 25 per cent of people are likely to have their sleep disturbed at an effective external level of about 56 dB(A).

Legislation also exists to limit the noise emission from new vehicles. This is set out in the Motor Vehicles (Construction and Use) Regulations,[2] which also describe the standard procedure for assessing compliance. These regulations have the effect of limiting the source noise level of traffic, and the values used tend to follow those set in the relevant Council of the European Communities directives. The problem remains, though, of ensuring that these limits continue to be met once the vehicle has been in use for a few years. No satisfactory, systematic means of easily assessing compliance has yet been devised, although in the case of motorcycles some progress has been made.[14]

Preventive measures

As with all environmental noise problems, there are three main areas where measures can be taken to reduce the noise level:

1   At source;
2   between the source and the receiver;
3   at the receiver.

At source

The noise emission limits discussed above are intended to encourage the design of quieter vehicles. There are, of course, economic constraints, because noise control techniques applied during the construction usually result in either increased production costs or increased running costs. As technological advances are made the cost of these techniques can be minimized and the gradual reduction in the noise limits that has occurred is an encouraging sign that progress is being made. The noise from lorries is a particular problem, although research has shown that using better exhaust systems could itself significantly reduce the noise emitted. There has also been successful research into producing a quiet heavy vehicle though the cost of implementing the techniques proposed has limited their adoption.[15] The time scale for the full benefits of these types of measures to materialize is very long, for even with the introduction of new, quieter vehicles all the older, noisier vehicles must first be phased out. However, because of the negative weighting given by dB(A) to low frequency noise the user of dB(A) may influence some vehicle manufacturers to concentrate on reducing noise emission in the mid-frequencies and, perhaps, neglecting the annoying lower frequency throb associated with HGVs.

The other treatment that falls into this category is traffic planning, that is, doing everything possible to keep the traffic as far away as possible from the people. For new roads this is just a matter of design, but there is inevitable environmental conflict. For example, from the traffic noise point of view the best solution for any new motorway is through the middle of green fields and parks. Again, a compromise must be sought, and this means making use of techniques that fall into the second category. The use of traffic management schemes can help with existing roads and, in particular, limiting the roads that heavy lorries may use can be effective.[16]

Between source and receiver

A significant reduction in noise level can be achieved by the use of a barrier if it introduces what is effectively an optical path difference between the source and the receiver.

Barriers can take the form of timber fences, seen alongside many motorways, concrete barriers or, more attractively, earth mounds which can be carefully designed to fit in with the landscape. Barriers must be heavy enough to be effective and as a rough guide:

1   for a barrier to give 0 to 10 dB(A) reduction, $M = 5\mathrm{kg/m^2}$;
2   for a barrier to give 10 to 15 dB(A) reduction, $M = 10\mathrm{kg/m^2}$;
3   for a barrier to give 15 to 20 dB(A) reduction, $M = 20 \mathrm{\ kg/m^2}$;

where M is the superficial mass of the barrier material.

So, if a reduction of 10 dB(A) is required, 12mm plywood at $6\mathrm{kg/m^2}$ would be adequate.[17] Barriers must, however, be continuous for any gaps in them will reduce their effectiveness.

The calculations used to estimate the reduction in noise level as a result of a barrier assume that the barrier is very long. In reality, this is virtually never the case, so it is good practice to compensate by allowing some tolerance during the barrier design calculations. It should also be noted that hard ground propagation is assumed when a barrier is included, for the sound path between source and receiver passes over the barrier and hence the attenuating effect of the ground is no longer relevant. The exception is a grass-covered earth mound rising directly from the receiver, which arguably should not be treated in the same way as a simple fence. (CRTN (1975), however, does not make this distinction.)

Planning to make use of the layout of the buildings on a development is a good way of minimizing the effect of traffic noise. In Figure 3.7(a) every dwelling and the courtyard have a sight of the road and hence are exposed to the noise. In Figure 3.7(b) more of the dwellings are shielded, and so is the courtyard.

At the receiver

The main treatment that falls into this category is ensuring that the facade insulation is adequate enough to enable internal guideline levels to be met (see 'Standard' above). Again, though, good use of internal layout can help. It is generally accepted that living rooms and bedrooms are noise sensitive whereas kitchens and bathrooms are not. Therefore in Figure 3.7(b), if the internal layout were arranged so that only kitchens and bathrooms overlook the road, the need for specific insulation measures for the other rooms would be reduced.

The insulation provided by a facade depends almost exclusively on the insulation of the windows. Their performance, in turn, depends on their size, whether they are single or double, the cavity size in a double set and how well they fit into their frames. Approximate insulation values are given in Table 3.5. The wide range in these results shows the variation that can occur as a consequence of, for example, the quality of the seals around the windows. In particular, the performance of the windows installed under the Noise Insulation Regulations is very dependent on the quality of the existing windows.

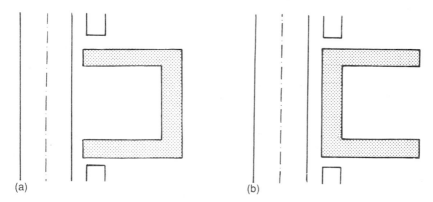

(a)   (b)

Figure 3.7(a)   *All dwellings have a sight of the road;* **(b)** *Some dwellings and all of the courtyard are shielded from the road*

Table 3.5

Approximate insulation values of windows

Type	Value
Open single window	5–15 dB(A)
Shut single window	10–25 dB(A)
'Staggered' opened double windows (100mm cavity)*	17–27 dB(A)
Shut double windows (100mm cavity)†	25–38 dB(A)
Shut double windows (50 mm cavity)‡	20–40 dB(A)

*Opened in a staggered manner such that there is no direct sound path through.
†An existing window upgraded to secondary glazing according to the Noise Insulation Regulations
‡Heavy glass must be used to give the highest performance.

Each element of the facade must in itself provide the required insulation so, for example, not only the windows but also any balcony doors must be treated. Balconies, in fact, can assist the insulation obtained because they can have a beneficial shielding effect. If the internal guidelines can be met only by having the windows shut, alternative ventilation should be provided, in the form of either electrically driven ventilators (as specified by the Noise Insulation Regulations) or static ventilators or air bricks. It is essential that they provide sufficient sound insulation otherwise the desired overall facade insulation will not be achieved. Specially designed acoustic ventilators and air bricks are available to meet this need. It is also important that the mechanical ventilator does not make too much noise when running. A limit is specified in the Noise Insulation Regulations but, even so, the operation of the ventilators can effectively reduce the facade insulation by 2 dB(A).[18]

Some research has been carried out into the effectiveness of the Noise Insulation Regulations package in terms of both the acoustic performance and the subjective reaction. One of the conclusions which emerged showed that mechanical ventilators were not very popular and were often not used.[19] What did become clear was that people liked the flexibility of being able to open and shut their windows as they wished – the feeling that they exercised some control over the noise source.

Vibration

Road traffic generates vibration which, when perceived, can cause both annoyance and a fear that buildings are being damaged.

Figure 3.8 shows the two main transmission paths for vibration. For path A vibration is generated as a result of the wheels of the vehicle travelling over the uneven road surface. It is then transmitted through the ground and

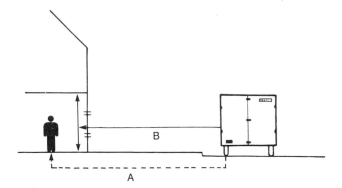

Figure 3.8   *Transmission paths for traffic induced vibration*

into the structure of the building. Path B arises from the low frequency noise generated by the vehicle which then excites the building (usually via the window) and is perceived as vibration.[20]

Guidance regarding the acceptable levels of vibration in buildings is given in BS 6472/1984, *Evaluation of Human Exposure to Vibration in Buildings (1 Hz–80Hz)*.[21] The level of vibration is measured in terms of either rms acceleration or peak velocity. As a general rule, vibration arising via path A usually occurs at a lower frequency (around 12 Hz) than the noise-induced vibration which is more prominent at higher frequencies (50–80 Hz). As implied by the title this standard considers the effects on people. There is, at present, no British standard with regard to the likely damage to buildings, and consequently a draft German standard, DIN 4150, tends to be used.[22] Guide values are given in terms of peak particle velocity and depend on the type of the building. Road traffic can be considered as generating steady state vibration and the guide value is a peak velocity of 5mm/s (the maximum value measured in any of three orthogonal directions on the top storey of the building). Not enough is known about this subject, however, because there are many cases where damage (usually in the form of plaster cracking on walls) seems to have occurred at far lower levels than that quoted in DIN 4150. Historic buildings are thought to be particularly at risk, not only because of their age, but also because they tend to be situated near busy roads.[23]

Work has been carried out both to relate levels of vibration to levels of noise and to assess people's reaction to various levels of vibration. It has been found that the subjective assessment of the vibration and the eighteen-hour L_{Aeq} noise level correlate well, confirming a close link between traffic noise and vibration.[24]

Some work has been carried out to try to be able to predict vibration levels from traffic flow. It is extremely difficult to do so because of the influence of both the ground soil conditions between the road and the building, and the type of building. Some success has been achieved in attempting to predict the change in vibration that will result from a change in traffic flow.

The most effective way of reducing the vibration generated from tyre/road interaction is to ensure that the road surface is smooth and even. It has also been thought that digging a trench between the road and the affected building would reduce this vibration, the theory being that the length of the transmission path is effectively extended by the trench. In practice the results have been disappointing and of course, in many situations, there are considerable practical problems regarding the location of such a trench.

The airborne induced vibration can be partly reduced by increasing the facade insulation. Secondary glazing, however, is not very effective at reducing the noise in the frequency range 50–80 Hz (typically 13–20 dB) and, almost inevitably, some of this noise will enter and excite the structure. The main method available is to reduce the low frequency noise output of the vehicle, either directly through the technological approaches mentioned above, or indirectly by routing heavy vehicles away from the affected buildings by the use of traffic management schemes or lorry bans.

Aircraft noise

Under section 79 of the Civil Aviation Act[25] the Department of Transport is empowered to introduce noise-reduction measures at designated aerodromes. It is under this section that the British Airports Authority operates on behalf of the Department of Transport a series of thirteen noise monitoring sites around Heathrow Airport. These sites are set under the departure routes and near to noise-sensitive areas. It is claimed that as there is far less scope for altering landing procedures there is no point in monitoring noise from landing aircraft. The local authorities near to Heathrow Airport do not subscribe to this view and have operated a monitoring scheme of both taking-off and landing aircraft for many years. The results are being used for planning purposes and for assessing long-term trends. The Department of Transport has set maximum noise limits at its thirteen sites of 110 PNdB for daytime and 102 PNdB for night-time. For serious or continued infringements of the limits airline operators may have penalties imposed on them, even so it is sometimes necessary to protect the affected residential areas by means of insulation.

Insulation

Insulation against aircraft noise presents particular problems not encountered with other sources. For a start the source is in the air and can affect all facades of a house, unlike road traffic which generally will affect only one. Although aircraft fly specified routes it is not unusual, particularly on take-off, for them to stray by half a mile or so from the nominal flight route. Also jet engines produce high levels of low frequency noise against which it is especially difficult to insulate.   A typical insulation package would involve acoustic double glazing with acoustic absorbent lining to the reveals. The two panes should be of different thickness with at least a 100 mm gap between them, preferably 150 mm. As insulation will be achieved only with the windows closed, an alternative method of ventilating the room must be provided. This is best done by providing a form of mechanical ventilator, which must be of the noise attenuating type so that it does not form an easy flanking transmission path. Care must also be taken to ensure that noise from the operation of the ventilator does not cause a nuisance. Another problem with double glazing is overheating due to solar gain, which can be avoided by fitting venetian blinds between the panes.

The roof must be treated against aircraft noise by increasing the mass of the top-floor ceilings. This can conveniently be done by introducing pugging to increase the ceiling mass by 10 kg/m^2. It should be noted that thermal insulation of lofts will have a negligible effect on the transmission of sound through the ceilings. Where fireplace openings are no longer used they should be blocked off with a partition of at least 10 kg/m^2 surface density, although care must be taken to provide adequate ventilation to the chimney. If an external door opens directly into a habitable room some form of lobby area may be necessary to achieve satisfactory insulation. In some areas grant schemes are available or have been available to give financial assistance to residents towards the cost of insulation against aircraft noise.

Planning

In the long term local authorities can minimize noise nuisance by careful planning. Advice was given to local authorities by the Department of the Environment in 1973 in the form of a circular entitled *Planning and Noise*.[11] Section 15 stresses the importance of controlling noise-sensitive developments near airports and refers to appendix 2, which contains criteria for the control of development in areas affected by aircraft noise. These criteria cover dwellings, schools, hospitals, offices and factories. Depending on the level of aircraft noise exposure now or expected within the next fifteen years, the criteria recommend whether planning permission should be refused, granted or granted on condition that noise insulation measures are provided. The noise levels are expressed in terms of the Noise and Number Index (NNI) (see below). For residential areas the circular recommends that for areas between 40 and 59 NNI no large new developments should be allowed but infilling can be allowed with appropriate insulation. It further recommends refusal of permission where exposure exceeds 60 NNI. The circular recognizes that these criteria may not apply in all circumstances and that some urban areas may require relaxation of the recommendations whereas in rural situations they may be more stringent.

Helicopter noise

Helicopter noise is different from fixed wing aircraft noise in many respects. Blade slap of the main rotor is a common phenomenon and can introduce an irritating impulsive character to the noise. However, a study carried out by the Civil Aviation Authority in 1983[26] concluded that despite its shortcomings NNI was still the best index for relating annoyance to helicopter noise exposure as well as for fixed wing aircraft. In practice this has proved suitable for helicopters flying on a specified route but not for assessing the effect of helicopter movements in the close environs of a heliport. Helicopters have the ability to hover, which means the duration of the event can be quite prolonged although the NNI formula will take no additional account of this. Also, due to the increased manoeuvrability of helicopters, heliports can be, and very often are, sited in city centres, where the potential for annoyance is far greater than for a conventional airport. It should also be noted that, provided safety requirements are met, any area can be used as a heliport without planning permission so long as it is not used on more than twenty-eight days in any one year. NNI therefore may not be a good predictor of annoyance near to a heliport. In recent years it has become common practice to estimate or measure noise levels in terms of the equivalent continuous noise level (L_{Aeq}) as well as NNI when assessing situations near to heliports. L_{Aeq} has the advantage that it inherently takes account of the duration of the event as well as the maximum level.

The GLC took the initiative some years ago to attempt to control noise from helicopters before it became a serious nuisance in London. The objective was to prevent any premises being exposed to helicopter noise in excess of 40 NNI. Planning authorities have no powers to control noise nuisance by helicopters (or fixed wing aircraft) in flight but they do have powers under a Town and Country Planning Act 1971, section 52 agreement to control the use of a heliport, and thus exercise some control over helicopters in flight. Using such an agreement the GLC and the planning authorities to whom its planning powers have subsequently devolved have limited both the number and type of helicopters permitted to use London's various heliports. The agreement made use of a list, referred to as list A, which named helicopters judged to be more acceptable on the grounds of environmental noise. Quotas were then set for the annual number of movements to be allowed in terms of list A and non-list A helicopters. The criterion to be met for inclusion on list A was that the helicopter should not produce more noise than 81 dB(A) at 150 metres at right-angles from the landing or take-off point.

The certification of helicopter noise was recommended by the International Civil Aviation Organization (ICAO) Committee on Aircraft Noise (CAN) in May 1983. The proposals classify helicopters by five categories depending on their all-up weight and their noise emission characteristics during take-off, landing and flyover. A paper by Simson and Turner[27] describes classification and provides a table showing how list A helicopters relate to the new categories for various helicopter types. It is envisaged that in future planning assessments the new ICAO categories will be used rather than the old list A classification, for two principal reasons. First, the ICAO test procedure is carried out under closely controlled flight operating conditions and hence should yield more reproduceable results and, second, the certification information is likely to become readily available for different helicopter types.

Noise and number index (NNI)

The index most commonly used in the United Kingdom is the NNI. It is based on the PNL and the number of aircraft passing the assessment point, and is calculated from the following formula

$$NNI = L + 15\lg N - 80$$

where L = logarithmic average maximum noise level in PNdB, and N = number of aircraft.

NNI takes its origin from the 1963 Wilson Committee report on noise.[1] The Committee commissioned a survey around Heathrow Airport, involving noise measurements at eight-five typical locations and administering a forty-two question survey to 1731 people affected by aircraft noise. A second survey, carried out in 1967[28] to re-examine and extend the original survey, confirmed the results of the earlier survey but did express some doubt about the validity of the 15 factor. The second survey more closely defined the number parameter N by specifying it to be the number of aircraft exceeding 80 PNdB during the twelve-hour period from 0600 to 1800 GMT averaged over the three-month summer period from mid-June to mid-September. It is worth pointing out that during this summer period British Summer Time is in operation, hence the NNI period is 0700 to 1900 BST.

The NNI calculated above is therefore a long-term average over the summer period, but on any one day, due to wind direction or operational procedures, the exposure of any particular point may be considerably worse. One suggestion is that people's reaction with respect to annoyance relates more closely to the worst exposure over a particular period rather than the average exposure, and this hypothesis was to some extent borne out by the second survey of aircraft noise, which showed slightly better correlation of annoyance with worst-mode NNI. Even so it has remained common practice to use average-mode NNI when assessing aircraft noise.

Points of equal NNI can be linked and drawn on a map to show NNI contours and each year the Civil Aviation Authority issues NNI contours for Heathrow, Gatwick and Stansted airports. Contours are normally drawn at 5 NNI intervals from 35 upwards where 35 NNI relates to low annoyance, 45 NNI to moderate annoyance and 55 NNI and above to high annoyance.

Equivalent continuous noise level, L_{eq}

Although NNI is currently the most common index used for the assessment of aircraft noise there is a recent trend towards using L_{Aeq}. L_{Aeq} has several advantages over NNI, the main one being that it can be measured directly whereas NNI requires complex calculations. The Department of Transport has recently commissioned the Civil Aviation Authority to conduct a survey either to substantiate the use of NNI or to devise a new index. The survey has been reported in full in a Civil Aviation Authority paper[29] and a brief summary reported by Brooker and Richmond.[30] Some 2097 people were interviewed covering areas around Heathrow, Gatwick, Luton, Manchester and Aberdeen airports. The levels of annoyance determined from the interviews were correlated with noise measurements carried out in twenty-six areas with at least twenty-two days' monitoring at each site. The main conclusions of the study criticized NNI to some extent, stating: 'Technically, on the study results, L_{Aeq}, measured over 24 hours, is a better descriptor of the disturbance responses than NNI.' It went on to give approximate equivalents to 35 and 55 NNI as being 56 and 70 L_{Aeq} and suggests that linear interpolation between these values will be adequate for most current airport environments.

Railway noise

The basic railway network in Great Britain was established by the beginning of this century and has remained largely unchanged except for some closures. In London alone there are approximately 500 British Rail route miles compared with 1000 miles of principal trunk roads. Interest in railway noise increased considerably in the 1970s for two main reasons: first, due to the original plans to build a channel tunnel with a rail link to central London and, second, due to the practice of local authorities using narrow strips of land alongside railway tracks for housing developments because of the chronic shortage of suitable sites. Since some housing is still being constructed close to railways and the channel tunnel plans include a possible high-speed link between a new terminus near the south coast and a new central London terminus, this has caused a renewed interest in noise annoyance from railway operations.

Sources

There are two main noise sources from trains, the locomotive noise and the wheel-rail interaction noise. Vibrating body panels can also become a noise source, although generally a problem only with empty freight stock. Another important noise source in certain particular locations is wheel squeal. This occurs only where the track forms tight curves and it is caused by sliding contact of the wheel flanges. It is a very obtrusive noise consisting of piercing high frequency tones. Fortunately the problem can be relatively easily solved by the use of automatic flange lubricators fixed at appropriate points on the track.

Locomotive noise

Noise from the locomotive is not usually a problem at normal operating speeds. Most complaints involving locomotive noise can be attributed to slow-moving trains where the power unit is under load, particularly with diesel locomotives.

Diesel engines tend to produce high levels of low frequency noise, which can cause particular problems of annoyance. A simple 'A' weighted measurement may not adequately predict annoyance because the 'A' weighting network takes little account of low frequencies. These high levels of low frequency sound can often cause vibration in lightweight parts of a building fabric such as windows, giving rise to annoyance and to the common misconception that the whole building is vibrating.

It is quite common to have more than one power unit in a single train, and these are termed multiple units. They may be electric multiple units (EMU) or diesel multiple units (DMU). DMUs can cause similar problems to those discussed above except that having several driving units per train means that each diesel unit can be smaller and will cause less noise. Noise from DMUs is clearly audible at the trackside for speeds up to 110km/hr when wheel-rail noise, which is speed dependent, will become the dominant noise source. EMUs are of two designs depending on the region. BR Southern Region uses an electrified third rail system whereas elsewhere an overhead cable system is used. In general, from a noise emission point of view, the overhead cable system is to be preferred. The third rail pick-up system seems to be consistently noisier and although the reasons for this are not known it is likely that the fairly massive current collection apparatus is the source of the additional noise. Even so EMU noise is not usually a problem, being indistinguishable from the wheel-rail noise at normal operating speeds.

Wheel-rail noise

Wheel-rail noise is produced by the rolling contact of the wheel on the rail. There has been some considerable investigation in the past to determine whether the wheel or the rail is the greater noise radiator and on balance the evidence points towards the wheel. The wheel vibrates in such a way as to approximate to a dipole noise source. The noise produced by the wheel-rail interface is heavily modified by various factors, one of which is whether the track is jointed or continuously welded rail (CWR). Jointed track is the traditional type of track made up of separate lengths of rail with expansion gaps between them. The rails are of standard lengths so that as the wheel passes over the discontinuities a rhythmic noise is produced. CWR has the joints filled with a malleable solder which can squeeze out to allow for expansion but which bridges the discontinuity, and reduces wayside noise levels by about 5 dB(A).

Track is usually laid on ballast and sleepers, which require occasional maintenance and cleaning. An alternative is concrete slab track, which is convenient to lay and is maintenance free, but unfortunately it is likely to increase noise levels by 2 to 4 dB(A).

Wheel-rail noise is further modified by the action of the train under braking. For older coaches (Mark I and II) noise levels can be expected to increase by 5 to 8 dB(A) while braking. The newer coaches (Mark III) use disc brakes rather than tread brakes, which produce less noise, hence the increase expected would be of the order of only 2 dB(A). In general the more modern coaches are 9 dB(A) quieter than earlier coaches, partly due to the use of disc brakes and partly due to the wheel design.

Corrugations in the rail head, or 'roaring rail' as it is sometimes referred to, is yet another factor determining the wheel-rail noise produced. Small variations of a few millimetres in dimension, which sometimes arise in the rail head, can cause dramatic increases in noise level from 2 to 7 dB(A) or even up to 15 dB(A) in severe cases. The reason for the formation of corrugations is not really understood but one possible suggestion is uneven wear due to variations in hardness of the metal when the rail was formed. The only solution is to grind the rail head to a smooth flat surface but this is both time consuming and expensive.

Prediction

The classic research into railway noise prediction was carried out by Peters.[31,32] The method produces a time history profile and is based on the assumption that the wheel-rail noise is a dipole source. Comparison of predicted and measured data shows very good agreement, generally within 3 dB. A simplified prediction method is given by Stanworth in *Railway Noise and the Environment*,[33] which contains a series of graphs allowing the maximum noise level to be estimated at a specified distance. Distance corrections can then be applied for the required distance. The document also quotes a few rules of thumb which can be summarized as follows. The maximum noise level will increase by 9 dB(A) for a doubling of speed. For distances of up to half a train length the maximum noise level will reduce by 3 dB(A) for a doubling of distance. Beyond one train length the reduction will increase to 6 dB(A) for a doubling of distance. Interpolation can be carried out for distances between half and one train length. It should be stated that these are rough estimations if actual noise measurements are not possible but in all cases measurements are to be preferred.

The original Noise Advisory Council published a booklet in 1978[34] which gives guidance on both the measurement and prediction of L_{Aeq} with a specific section concerning railway noise. A predictive method is described based on the maximum noise level of the locomotive and the maximum noise level from the wheel-rail using L_{AX} the single event noise exposure level. The formulae used in the method are based on those of Peters.

The L_{AX} for the locomotive can be estimated from

$$L_{AX1} = L_{Amax} + 10\lg(d/v) + 8.6 \qquad dB(A)$$

where d = perpendicular distance to track in metres
v = train speed in km/hr
L_{Amax} = maximum sound pressure level of the locomotive in dB(A).

The L_{AX} for the wheel-rail noise can be estimated from

$$L_{AX2} = L_{Amax} + 10\lg(Lt/v) + C \qquad dB(A)$$

where L_{Amax} = maximum sound pressure level of the wheel rail noise in dB(A)
Lt = train length in metres
v = train speed in km/hr
C = constant.

The constant C can easily be obtained using the graphical form published in the Noise Advisory Council booklet.[34]
The two L_{AX} values can then be combined to give a total L_{AX} using the following formula

$$L_{AX} = 10\lg(10^{L_{AX1}/10} + 10^{L_{AX2}/10}) \qquad dB(A)$$

Having obtained L_{AX} values for each train type then the L_{eq} for time T seconds can be calculated using

$$L_{Aeq} = 10\lg 1/T \int_{i=1}^{n} 10^{L_{AXi}10} \qquad dB(A)$$

Criteria

There is no national legislation or guidance in the UK giving criteria for the acceptability of railway noise. Circular 10/73 *Planning and Noise*[11] makes no mention of railway noise. It was because of this lack of guidance that the GLC decided in 1976 to issue a set of guidelines[12] covering railway noise and residential developments. After some research and after comparing annoyance by railway noise with the annoyance caused by an eighteen-hour L_{A10} of 68 dB(A) for road traffic noise, it was decided to adopt a criteria of 65 dB(A) as a twenty-four-hour L_{Aeq}. If external noise measured one metre from the facade exceeded this level some form of ameliorative treatment should be used. If levels exceeded 80 dB(A) compulsory purchase should be sought.

Noise reduction

The best noise reduction technique is to reduce noise levels at source. This is obviously a very long-term process but British Rail is having some success in this respect as the Mark III rolling stock is considerably quieter than its predecessors.

Lineside barriers are a possible method of reducing noise. They have the advantage that the main source, the wheel-rail noise, is close to the ground so that barriers do not need to be very high to give good attenuation. They should be situated as near to the source as possible and should at least prevent a clear line of sight from the receiver location to the rail. The disadvantage of lineside barriers is that where there is more than a single line it is difficult to shield the more distant tracks adequately. British Rail is also reluctant to install barriers close to the track due to the safety aspects for track maintenance workers.

As a last resort the receiver location can be protected by use of a sound insulation package such as double glazing with a noise attenuating mechanical ventilator. It is good practice when designing double glazing to use glass of different thicknesses with a gap between the panes of 100 to 150 mm. The reveals should be lined with acoustically absorbent material. In cases where high levels of low frequency noise are expected, say from slow-moving diesel locomotives, extra care should be given to the design and an air gap of at least 150mm between the panes should be used.

Annoyance

In general terms people appear to be more tolerant of railway noise by the equivalent of 5 dB(A). There are various possible reasons for this: the railway has always been there, annoyed individuals do not know to whom they should complain and railways are a public form of transport available to anybody – unlike cars, which are a personalized form of transport causing some feelings of resentment. This increased tolerance is, of course, generally not found where a new railway line is constructed near to existing housing or even where new housing is constructed near an existing railway line. As an example, a new light railway system began regular passenger operations in London's Docklands in 1987 and although some of the route follows alongside existing BR tracks other parts follow a completely new elevated line. Complaints immediately arose concerning noise from its operation but it is not clear whether or not the local residents have an increased sensitivity to the new railway operation introduced into their locality. This particular situation is further complicated as the entire route is elevated, which may account for the reported high levels of low frequency noise. There also seems to be some concern about high frequency noise from the airbrake system.

Fields and Walker confirmed this general tendency to accept railway noise in the results of their survey of railway noise in Great Britain.[35] This investigation involved 2000 noise measurements at 403 different locations, combined with a questionnaire survey of 1453 respondents. The main conclusions of the report were, first, that roughly 2 per cent of the population of England are bothered by railway noise and that something like 170 000 people in Great Britain live at railway noise levels in excess of a twenty-four-hour L_{Aeq} of 65 dB(A). They confirmed that twenty-four-hour L_{Aeq} was the best noise index currently available for assessing railway noise. The relationship between noise and annoyance was basically linear for levels above 45 L_{Aeq} with no particular level of onset of annoyance. They agreed with previous surveys that overhead electric trains caused less annoyance than other types. One of the more surprising conclusions was that noise from maintenance operations was rated as being more annoying than train movements.

Construction noise

Construction and demolition noise is frequently the cause of complaints to local authorities, particularly in large conurbations where suitable land for building is at a premium. Construction and demolition work often entails very noisy processes such as the use of pneumatic drills or pile driving, sometimes for considerable lengths of time as in the case of road construction. Local authorities have powers under two very important Acts of Parliament which enable them to exercise control over noise and to provide compensation for householders adversely affected by construction and demolition noise.

The Land Compensation Act

The Land Compensation Act 1973[3] provides the legal machinery with which to compensate people whose enjoyment of their property has been adversely affected by the execution of public works. Under this Act regulations can be introduced as guidelines and to provide working rules to assess whether compensation is due. Only one such set of regulations has so far been introduced, the Noise Insulation Regulations 1975,[4] which superseded the earlier 1973 regulations. These regulations apply only to new highways, and provide a sound insulation package for qualifying buildings that will be adversely affected by road traffic noise. They also make provisions for dwellings that are or will be adversely affected by noise caused by the construction stage of the work irrespective of whether the property would or would not qualify on future predicted road traffic noise levels. An anomaly of the regulations is that although the qualifying criteria are very specifically laid down for road traffic noise caused once the road is built, the qualifying criteria for properties affected during the construction stage are left very much open to interpretation. The regulations merely say that an offer of compensation should be made to those who in the opinion of the highway authority are 'seriously affected' for a 'substantial period of time'. No guidance on the interpretation of these criteria has been issued by central government but a set of guidelines was developed by the GLC[12] so that it could be seen to be fair and consistent in its approach. The criterion level was set at an L_{Aeq} of 75 dB(A) for the period from 0700 to 1900 hrs, reducing to 65 dB(A) for 1900 to 2200 hrs and reducing again to 55 dB(A) for the remaining period from 2200 to 0700 hrs. The qualifying clause, 'a substantial period of time,' was interpreted as a period of ten working days. If both the above criteria were expected to be exceeded the same sound insulating package specified for road traffic noise was offered to ameliorate the nuisance.

The Control of Pollution Act

Only criteria for compensation have been discussed so far in this chapter and these criteria are not necessarily suitable for the control of noise. Local authorities have been given wide-ranging powers of control under the Control of Pollution Act 1974.[36] Part III of this Act deals with noise and sections 60 and 61 deal specifically with construction noise. Section 60 empowers the local authority to publish a notice when it believes work is being or is going to be carried out which may cause a nuisance. The notice may specify particular items of plant to be used or items which must not be used, and also the allowed hours of working and noise emission levels. When acting under section 60 the local authority must have due regard for

1   the 'best practicable means' to prevent the nuisance;
2   equally effective alternative measures;
3   the need to protect the locality, and
4   any relevant code of practice in existence.

A code of practice is in existence in the form of BS 5228, which will be discussed later in more detail. An appeals mechanism exists for the agents on whom the notice has been served if they feel the terms of the notice are unfairly restrictive.

Section 61 of the Act is often referred to as the 'prior consent' section. Under this section a developer may require the local authority to make known its noise requirements before the work starts. After the developer submits all the relevant details of the type of work to be carried out, the local authority must assess the situation and either give or refuse its consent for the work to be carried out, within twenty-eight days of the application. It may qualify its consent by attaching conditions. If the authority fails to respond in the time period or if the developer objects to conditions, once again an appeals procedure exists. Having gained consent under section 61 the developer is then immune from any actions for nuisance under section 60.

British Standard 5228

As both sections 60 and 61 can be pre-emptive, prediction methods are essential. The best prediction method available is contained in a British Standard BS 5228 *Noise Control on Construction and Open Sites*.[37] The standard is divided into four. Part 1 describes the general prediction method and gives basic information and procedures for noise control. Part 2 describes the legislation applicable to construction and demolition sites. Part 3 deals specifically with surface coal extraction by opencast methods and draft Part 4 describes noise control methods applicable to piling operations. It is the intention of the British Standards Institute to issue further parts to this standard as and when required by the industry. BS 5228 does not cover vibration even though the Control of Pollution Act defines noise as including vibration; instead the standard directs that BS 6472,[21] which covers human response to vibration in buildings, should be used.

BS 5228 is concerned with both the protection of persons on site from noise-induced hearing loss and the protection of people in the neighbourhood from annoyance. Advice is given on the criteria to be used in setting noise control targets but no overall limit is given as this will obviously depend on each individual situation. For example, on a construction site in an area consisting of offices with no nearby residential properties, noisy operations will cause less annoyance if they are carried out at night rather than during the day. The standard gives noise levels likely to be produced by various items of plant encountered on construction sites. Each item is classified according to its size, in terms of weight or power, and according to the type of use, for example demolition, concreting and so on. For each activity a typical activity L_{Aeq} is given. The activity L_{Aeq} is the 'A' weighted L_{Aeq} noise level measured at 10m from the source for one complete cycle of an activity. By making appropriate distance and screening corrections and by analysing all events expected to be carried out on site an estimate of the L_{Aeq} can be made at any particular receiving point. As an alternative method predictions can be based on typical sound power levels and percentage 'on times', values of which are also tabulated in the standard.

Noise from pile driving is a particularly difficult case and justifies its own part of the British Standard. Once again tables of data are given for varying pile types and size and for various piling methods. Pile driving methods are very dependent on the soil type, and for engineering reasons it may not always be possible to substitute a quieter method. Drop hammer piling can be quietened by introducing some form of fibrous cushioning material between the pile head and the hammer, but this is likely to affect the efficiency of driving of the pile and hence cause the activity to last longer. A better method is to use a hush piling rig, which consists of a suspended enclosure that completely surrounds the pile and the driving hammer. This technique can reduce maximum noise levels by 25 dB(A).

Noise reduction

Screening is a useful technique that can give noise level reductions of 5 to 10 dB(A) or even up to 15 dB(A) for well-constructed purpose-built screens. As a general rule, if the source is just obscured from sight by the screen a reduction of about 5 dB(A) can be expected, with an increasing attenuation for higher or longer screens. The on-site storage of materials such as bricks can often by used as a very effective temporary barrier for noisy processes. Complete enclosure will generally give better results, of the order of 15 to 20 dB(A), although due attention must be paid to the ventilation requirements of the enclosed plant. As a longer-term measure silencing treatment can often reduce plant noise levels by 5 to 10 dB(A) by the use of more efficient silencers or by enclosures. One such example is the air compressor where the newer style enclosed versions are some 15 dB(A) quieter than the old versions. More emphasis will in future be placed on the development of quieter plant as EEC legislation to control noise from construction plant takes effect.

As a general rule most site operations can easily be carried out in a quieter way if the site employees are aware of the possible nuisance they may create.

Industrial noise

A whole range of different sounding types of noise from a wide variety of sources fall into this category. It might be a hum from a large chemical processing plant or bangs and crashes from a small scrapyard. Whatever the noise the environmental problem arises when these businesses are located close to residential areas or vice versa. From the previous discussion of noise rating, it will be appreciated that finding a suitable measurement parameter that estimates the subjective rating of all these different noises is not easy. These days, the equivalent continuous noise level (L_{Aeq}) would probably be used, but twenty years ago it had barely been developed and there were certainly no simple meters available to measure it. Yet, at that time, there was a great need to have some means of assessing the likely public reaction to a proposed new factory or to a new, noisier process commencing in an existing factory.

Wilson[1] commented on the widespread annoyance caused by industrial noise and in his report the results of work carried out by the Building Research Station to develop a simple assessment method are given. This procedure became adopted as a British Standard, and BS 4142, *Method of Rating Industrial Noise Affecting Mixed Residential and Industrial Areas* was published in 1967.[38]

BS 4142

This standard describes the measurement of industrial noise and how to predict whether that noise would be expected to give rise to complaints. It assumes that complaints about a noise are likely when the level of that noise exceeds the background level by a certain amount. The standard does not, however, attempt to deal with assessing annoyance, and it is most important that a clear distinction is maintained between annoyance and complaints. Generally annoyance about noise will occur first, and the individual will be moved to complain only when that annoyance has reached a certain level. Not everyone, though, will complain at the same level of annoyance (some may never complain despite being very annoyed) so care is always needed when drawing conclusions about community attitudes based on complaints alone.

BS 4142 covers noise from factories, industrial premises and other fixed installations and is designed to be used for rating externally measured noise only. Using it for internal noise or noise generated from within the building is specifically excluded.

In predicting the likelihood of complaints the standard assumes that it is the industrial noise that changes. This means that it is able to assist in assessing the effect (in terms of complaints) of a new factory starting to operate near to a residential area, but it cannot assist in determining whether building new houses close to an existing factory is environmentally suitable.

Measuring factory noise according to BS 4142

The factory noise must be measured in dB(A) with the sound level meter set at rms (slow). The microphone position must be 1.2m high and at least 3.6m from any reflecting surface such as a house wall. The measurements should be made on the side of the houses nearest to the factory at a suitable location near to the houses from where complaints have arisen or may arise.

If the noise is fairly constant in level (that is, does not fluctuate by more than 10 dB(A)) the level is averaged visually, and this gives a value, L_s. If a higher level occurs from time to time as a result of a particular process, this level is also measured (L_h). The number of occurrences of this higher level and the duration of each are also noted. From tables and charts within the standard, corrections are made if the noise is tonal (that is, it has a clearly perceptible whine, hiss, hum or the like) or impulsive in quality (that is, there are bangs, crashes, and so on) or it is regular enough to attract attention. There is also an intermittency/duration correction which considers the percentage on time of the noise and distinguishes between day (0700–2200) and night. From this, the corrected noise level (CNL) is found (L'_s and L'_h). The CNL is then compared with the background noise, which is found by one of two methods.

Background noise level

BS 4142 states that the background noise level should preferably be measured and it is described by the statistical percentile L_{A90} (the level exceeded for 90 per cent of the time). The standard provides an alternative to measurement by describing a method to derive a notional background level. This is found by correcting a level of 50 dB(A) according to the type of installation (that is, whether it is a new factory or one that is being altered), the type of district (rural, surburban or industrial) and the time of day. By using this method the notional background level can vary from 40 dB(A) (for a new factory operating at night in a rural residential area) to 85 dB(A) (for an old established factory operating in an industrial area during the day). The notional background noise level has been subject to many criticisms and few now regard it as reliable.

Assessing likelihood of complaints

BS 4142 rates the noise in the following manner:

1 CNL − Background > 10 dB(A) – complaints likely;
2 CNL − Background = 5 dB(A) – marginal significance;
3 Background − CNL > 10 dB(A) – complaints unlikely.

CNL here is either L'_s or L'_h or both.

The problems of BS 4142

Although carefully developed in the early 1960s, BS 4142 has been beset with problems and now is generally not regarded very highly. Some of the criticisms include:

1 The notional background level can give very inaccurate estimates of the true background level. Errors of 20 dB(A) have been reported.[39] In general, the notional background level overestimates the true value, and hence the likelihood of complaints is underestimated.
2 Even measuring L_{A90} presents difficulties as the level is not constant over the twenty-four-hour period. Consequently, which value should be used – the maximum, the minimum, or a daytime hourly average?
3 The correction for intermittency of the factory noise reduces the level compared with a continuous noise. It has been argued that complaints may, in fact, be more likely with an intermittent noise.
4 BS 4142 tends to be used for assessing annoyance (which is beyond its scope) simply because there are no other standards available. In addition it has been applied to noise sources other than factories and, consequently, difficulties have arisen. Examples include noise from speedway tracks, noise from schoolchildren in a playground and even vibration from industrial premises.[40]

The shortcomings of this standard and the way it is used are well recognized, and in 1987 work commenced on redrafting it. Until a new standard is completed, however, BS 4142 will remain the only method available. As it is a British Standard, any legal action that may arise from complaints about noise from factories will have to place considerable emphasis on the standard's assessment. As it bears the BS seal of approval, the court have no alternative but to rely on it.

Scope for improvement

As indicated earlier, the use of L_{Aeq} rather than either L_{A90} or simple visual averaging of an rms signal would remove the need for many correction factors. L_{Aeq} reflects both level and duration and can now be easily and directly measured. It was thought that this would become the universal noise parameter so that regardless of the source the L_{Aeq} alone would be measured and an assessment made. That stage has not yet been reached, but it is gradually being used in an increasing number of situations.

Assessing the noise within the complainant's dwelling would also overcome some of the problems. First, it would seem more relevant to complainants, who would at least see the acoustician measuring the noise at the location where it is heard (that is, in their home). Use could also be made of noise rating and noise criteria curves, which look at the frequency content of the signal. Although this would mean a simple 'A' weighting assessment would no longer be adequate, it may be found that the correlation between the NC/NR measurement and the subjective response is better. There would also be more consistency in approach as NC/NR assessments are used in the case of noise from mechanical services.

Using the concept of increase in level over the background is still valid but with the ready availability of sound level meters, all background levels should now be measured and not estimated. This would at least mean that the basis of the assessment is more accurate. Using the notional background method has led to some ludicrous results in the past.

There is clearly much about BS 4142 that can be improved but, given the variety of noise that industry can create, let alone the vastly different responses of people to that noise, it will not be easy to develop a method that works well in all situations.

Draft BS 442 May 1988

The draft revision of BS4142 was published in May 1988 and many of the changes proposed above were included. The draft proposes the use of L_{Aeq} for measuring industrial noise which, of course, has the great benefit of eliminating duration corrections. This L_{Aeq} level is then compared with a *measured* L_{A90} background noise level and the difference is taken to be an indication of the likelihood of complaints.

The other major changes included in the draft are the elimination of notional background level, the elimination of the 5dB(A) marginal significance level, and the inclusion of the facility to measure the noise 1m from the facade at each relevant floor of the building and not only at ground level. The correction factors of 5dB(A) for noise with a distinct tonal or impulsive character remain.

The above changes have generally been welcomed even though the 5dB(A) correction factor for low-frequency tones may be a little on the conservative side. Especially welcomed by many concerned with low-frequency noise problems is the explicit reference in the draft that when 'the specific noise contains significant low-frequency components the procedure for assessing the noise from measurements of the 'A' weighted sound level may not be sufficient to determine the likelihood of complaints'.

Entertainment noise

During the last hundred years, recreation has become noisier. At the end of the nineteenth century virtually the only reportedly noisy pastimes were game shooting and, to a lesser extent, singing in public houses. Nowadays there are pop concerts, discotheques, public houses with live music, television, hi-fi, motor sports, water sports, model aircraft, do-it-yourself and clay pigeon shooting. With all these entertainments the balance must be struck between the rights of people to enjoy themselves and the rights of individuals not to be disturbed in their homes. One of the functions of the Control of Pollution Act 1974[36] is to provide a mechanism for controlling activities which generate noise that creates a nuisance to others; and noise must now also be a consideration when local authorities grant licences for entertainments, whether for large-scale pop concerts or for music and dance in the local public house.

In this section pop concerts and discotheque noise will be considered in some detail, primarily because most work has been carried out in those areas.

Pop concerts

Audience of up to 120 000 are entertained at a single pop concert. In the UK crowds of about 30 000 can be found at concerts held at most football grounds, around 70 000 at those held at Wembley Stadium, London and 120 000 at the Knebworth Park concerts in Hertfordshire.

In twenty years, since the mid-1960s, pop concerts have become synonymous with noise, and the amount of amplification available now seems virtually unlimited. It is not surprising therefore that environmental noise problems occur when these events are held.

The GLC in its *Code of Practice for Pop Concerts*[41] set down guidelines for exercising control over the noise from these events. Two aspects were considered: minimizing the risk of hearing damage to the audience, and minimizing the annoyance to those living nearby. The code soon became used not only in London, but also in other parts of the country and further afield. The control was set out in the form of a code of practice because it was felt that seeking co-operation with those running the concert was more effective than seeking pure enforcement. To minimize the risk of hearing damage, the code states that: 'The L_{Aeq} noise level at 50 metres from the speakers should not exceed the equivalent of 93 dB(A) for eight hours.'

The usual time/energy relationship operates so that for any concert duration, the L_{Aeq} limit can be found:

Time (hours)	L_{Aeq}
8	93
6	94
4	96
2	99

The code states that to minimize the annoyance to occupiers near the site at which an outdoor concert is held on no more than three days per year:

> The L_{Aeq} noise level measured for any fifteen-minute period of the concert or rehearsals outside the windows of their premises should not exceed the L_{Aeq} noise level measured during a comparable period when no pop concert or rehearsal is in progress by more than 10 dB(A) between 0700 and 2000 hours, and by more than 6 dB(A) between 2000 and 2300 hours. No sound from the premises should be audible within any other premises between 2300 and 0700 hours.

For concerts held on more than three days a year, the increase is limited to 1 dB(A) between 0700 and 2300 hours.

It is worth considering the various elements of this criterion.

1 The parameter to be measured is L_{Aeq}
2 The concept of an increase in level over a background level is used. The same comments made with regard to industrial noise apply here also, but in this case the background level is always measured.
3 The change in level at 2000 hours was included partly to reflect the increasing sensitivity of people as the evening passes, but partly to discourage promoters from continuing their concerts for too long. In practice it is not possible to change the limits at that time if it happens to be in the middle of the concert and, in fact, it has been found that the 10 dB(A) can apply throughout a concert as long as it ends no later than 2300 hours.[42]

    The use of 10 dB(A) as the permissible increase appears also to be reasonable when considered in terms of complaints. In Figure 3.9, it can seen that only one to two complaints tend to occur when the increase is no more than 10 dB(A), but once the level goes up by more than 13 dB(A) the number of complaints rises significantly.

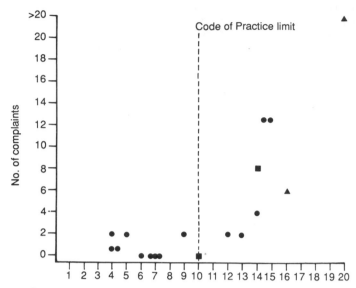

Figure 3.9   *Number of complaints with change of AL$_{Aeq}$ noise level from pop concerts*

4   The number of events per year is specified to differentiate between the occasional concert and the regular event. The number 'three', though, has no special significance and as long as the noise from the concerts meets the limits, it has been found that up to eight concerts can be held with no significant change in community response.[43]

For a particular venue, the Code of Practice provides the criteria that can be used, but the specific noise limits that must operate to meet the criteria are site-dependent. For the environmental aspect, the factors that will affect the limits include the background noise level and the sound attenuation between the concert arena and the housing. The attenuation can be increased by 4–5 dB(A) merely by angling the loudspeakers down towards the audience,[42] but otherwise it depends on the type of venue and the distance to the nearest housing. A typical attenuation for a football ground is 30 dB(A).

A procedure has now been developed which enables the noise level generated by a concert to be controlled effectively during the event and yet achieve the right balance between the requirements of the audience and of those living nearby. It has been found that by adopting this systematic approach, concerts can be held at venues which at first sight might seem environmentally unsuitable.[44]

Discotheques

In some respects the environmental problem of noise from discotheques is easier to solve because of the inherent increased sound attentuation that is available from the fabric of the building in which they are held. On the other hand, discos usually operate most nights of the week and can continue into the early hours of the morning. In addition, the potential risk of hearing damage is greater because an individual may choose to go to a disco virtually every night and, consequently, be regularly exposed to what can be high noise levels.

The first guidance about noise from discos again came from the GLC. In *Disco Rules OK?* criteria similar to those used for pop concerts were adopted.[45] For hearing damage risk, the L_{Aeq} noise level must not exceed an equivalent of 90 dB(A) for a disco that is open for eight hours. For the environment, the fifteen-minute L_{Aeq} must not increase over the background level for the comparable period. Inaudibility within the neighbouring premises again applies between 2300 and 0700. When comparing these limits with the Pop Code, the more stringent hearing damage risk criteria reflect the differing attendance patterns and hence exposure for the two types of event.

The environmental criterion is much more stringent to reflect the fact that discos operate virtually daily. This limit must be met by a combination of controlling the levels generated in the disco and by the fabric of the building providing adequate insulation. Clearly the hearing damage risk criterion also acts as a control of the level and this aspect has been further strengthened by the *Draft Code of Practice on Sound Levels in Discotheques* published by HMSO in 1986.[46] It is based on extensive work carried out by Bickerdike and Gregory, who investigated the noise exposure and attendance patterns

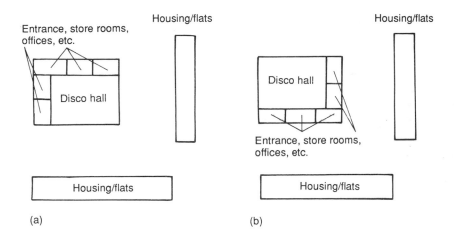

Figure 3.10 (a) *Poor internal layout;* **(b)** *Good internal layout*

of many people who go to discos.[47] From this, the concept of *maximum permissible exposure level* (MPEL) was developed. This is the maximum L_{Aeq} to which any member of the public can be exposed during the opening hours of a disco. The code states that the MPEL should not exceed 100 dB(A) for discos that have quieter rest areas (for example a bar area) or 95 dB(A) for those without. The noise level in the rest area should not exceed 85 dB(A).

The code gives advice on methods of achieving these limits whilst allowing the disco to be a satisfactory form of entertainment.

Even in many city areas, the background noise level in the evening could fall to about 45 dB(A), which means that for the environmental criteria to be met approximately 50 dB insulation is required. A solid brick or concrete structure will alone provide this degree of insulation, but as soon as doors, windows and roof lights are included, difficulties arise. For an existing structure a lobby system at the door and secondary or triple glazing with large air gaps are required. In addition mechanical ventilation/air conditioning would have to be provided as the windows would have to remain closed. For new buildings it is important to select a suitable location away from housing and, once chosen, to ensure that the layout within the building provides the optimum attenuation (Figure 3.10).

With many more public houses now operating discos, there are considerable problems when these premises are an integral part of a residential building. The noise is transmitted though the structure which means that the external L_{Aeq} increase is no longer an appropriate method of assessment. As yet there is no simple answer to this problem, although one licensing authority has found a solution by extending the inaudibility principle to cover all the opening hours of the pub.[48] It is arguable whether this is the fairest approach.

Model aircraft/ice cream chimes/audible intruder alarms

The Control of Pollution Act made provisions for codes of practice to be published to set out noise limits for specific items. In 1982 three were published, covering model aircraft, ice cream chimes and audible intruder alarms. These codes give noise limits and control other aspects of their operation. In the case of model aircraft, it is particularly ironic for those living near the major airports that action can be taken about noise from a model aircraft, but not from a real one. These codes do not have the force of law, but undoubtedly would be used by the courts in the case of any dispute.[49-51]

Clay pigeon shooting

This is a growing sport, and consequently, an increasing environmental noise problem. Again, the difficulty comes in trying to quantify the noise so that a judgement can be made regarding whether or not there is a nuisance. In addition to the noise level, the other factors affecting the response that so far have been identified are the number of shooting events in a year and the total number of shots fired. In 1987 a code of practice was in preparation which will, in due course, be published under the auspices of the Control of Pollution Act. It is likely to cover the various aspects that need to be considered to minimize the noise nuisance from this activity.

Motor sports

The Royal Automobile Club exercises close control over the noise levels from vehicles which participate in the various sports including autocross, cart racing and rallies. The club has a team of inspectors who ensure that the vehicles meet the required standards.[52]

Conclusion

Having considered some of the variety of noisy entertainments that exist, it will be seen that despite their differences there are some common factors and in the approach to controlling the noise from them, there are three common elements:

Finding the correct balance

This is the key to successful entertainment noise control in every case. On the one hand, a proposed activity at a certain location may at first seem environmentally unsuitable. Before preventing it, possible control methods must be considered. Conversely, allowing it to proceed without restriction may cause a nuisance and sufficiently prejudice the local community against the event that they will not wish it to take place again despite the promise of greater control.

Keep the criterion as simple as possible

Obtaining an objective assessment of the subjective response to an event can lead to complex equations and concepts which many people will not understand. Translating that into a criterion which states, for example, that a particular number should not be exceeded, will make objectives quite clear even though the number itself (it may be a one-minute L_{Aeq} or a noise rating) may not be understood. Acoustic principles and techniques should be used to give as straightforward a presentation as possible to the public of something that can be quite complex.

Maintain flexibility

All criteria have been set down to provide guidance. They are not necessarily the correct answer for each situation although some have been shown, over a period of time, to operate well in most circumstances. If it seems appropriate to do so, a different criterion should be adopted, but the justification for the change must be sound (especially if it means a tightening of the restriction). Flexibility in this way will help the right balance to be found.

References

1   *Noise – Final Report from the Committee on the Problem of Noise* (Chairman Sir Alan Wilson) (1963) Cmnd 2056, HMSO, London.
2   Motor Vehicle (Construction and Use) Regulations 1968 and subsequent amendments (1978) HMSO, London.
3   The Land Compensation Act 1973, HMSO, London.
4   The Noise Insulation Regulations 1975, HMSO, London.
5   Department of the Environment (1985) *Digest of Environmental Protection and Water Statistics*, HMSO, London.
6   Langdon, F.J. and Buller, I.B. (1977) *The Effects of Road Traffic Noise in residential Areas*, Building Research Station CP10/77.
7   Banks, R. (1974) *A Comparison of Traffic Noise in Wet and Dry Conditions*, Report DG/SSB/ESG/TM1, GLC.
8   Langdon, F.J. and Scholes, W. E. (1968) *The Traffic Noise Index: A Method of Controlling Noise Nuisance*, Building Research Station CP38/68.
9   Forsdyke, M.R. and Johnson, S, (1986) *The Comparison of L_{10} (18 hour) with other L_{10} Periods*, Report LSS/NV/TM142, London Scientific Services.
10   Department of the Environment (1975) *Calculation of Road Traffic Noise*, HMSO, London.
11   Department of the Environment (1973) *Planning and Noise*, Circular 10/73, HMSO, London.
12   Greater London Council (1976) *Guidelines for Environmental Noise and Vibration*, GLC.
13   BS 8233 (1987) *British Standard Code of Practice for Sound Insulation and Noise Reduction for Buildings*, British Standard Institute.
14   Lattimore, P.J. (1985) *Proposed Changes in EEC Legislation to control Motorcycle Noise Nuisance*, Report DG/SSB/ESD/R187, GLC.
15   Nelson, P.M. and Underwood, M.C.P. (1982) *Operational Performance of the TRRL Quiet Heavy Vehicle*, Transport and Road Research Laboratory Supplementary Report 746.
16   Lattimore, P.J. (1985) Lorry Noise Nuisance, *London Environmental Bulletin*, Vol. 3, No. 1.
17   Prediction of Road Traffic Noise Part 2 (1976) *Building Research Digest*, 186.
18   Turner, S.W. London Scientific Services Report LSS/LWMP/S6. (1988), *Traffic Noise and Secondary Glazing*, LSS.
19   Greater London Council (1984) *Noise Annoyance in Homes with Secondary Glazing – Part II The Social Survey*, GLC.
20   Martin, D.J., Nelson, P.M. and Hill, R.C. (1978) *Measurement and*

Analysis of Traffic-induced Vibrations in Buildings, Transport and Road Research Laboratory Supplementary Report 402.

21 BS 6472 (1984) *Guide to the Evaluation of Human Exposure to Vibration in Buildings (1 Hz–80 Hz)*, British Standards Institute.

22 DIN 4150 (Part 3) (1983) *Vibrations in Buildings, Effects on Structures*, Draft Standard, Deutsches Institut fur Normung eV.

23 Ellis, P. (1985) The Effects of Traffic-Induced Vibration on Historic Buildings, *Proc. IOA*, Vol. 7, Part 2.

24 Watts, G.R. (1984) *Vibrational Nuisance from Road traffic – Results of a 50 Site Survey*, Transport and Road Research Laboratory Report 1119.

25 Civil Aviation Act 1982, HMSO, London.

26 Atkins, C.L.R., Brooker, P. and Critchley, J.B. (1983) *1982 Helicopter Disturbance Study: Main Report*, Civil Aviation Authority, DR Report 8404.

27 Simson, J. and Turner, S. W. (1985) Helicopter Noise Classification, *London Environmental Bulletin*, Vol. 3, No. 1.

28 *Second Survey of Aircraft Annoyance Around London (Heathrow) Airport* (1971) HMSO, London.

29 Brooker, P., Critchley, J.B., Monkman, D.J. and Richmond, C. (1986) *United Kingdom Aircraft Noise Study: Main Report*, Civil Aviation Authority, DR Report 8402.

30 Brooker, P. and Richmond, C. (1985) The United Kingdom Aircraft Noise Index Study Part 1 – Main results, *Proc. IOA*, Vol. 7, Part 2.

31 Peters, S. (1973) Prediction of Rail–Wheel Noise from High Speed Trains, *Acustica*, Vol. 28, pp. 318–21.

32 Peters, S. (1974) The Prediction of Railway Noise Profiles, *Journal of Sound and Vibration*, Vol. 32, No. 1, pp. 87–99.

33 Stanworth, C.G. (1977) *Railway Noise and the Environment – A summary*, 2nd edn, TN PHYS 4, British Railways Board, London.

34 The Noise Advisory Council (1978) *A Guide to the Measurement and Prediction of the Equivalent Sound Level L_{eq}*, HMSO, London.

35 Fields, J.M. and Walker, J.G. (1982) The Response to Railway Noise in Residential Areas in Great Britain, *Journal of Sound and Vibration*, Vol. 85, No. 2, pp. 177–255.

36 The Control of Pollution Act 1974 HMSO, London.

37 BS 5228 (1987) *Noise Control on Construction and Open Sites*, British Standards Institute.

38 BS 4142 (1967) *Method of Rating Industrial Noise Affecting Mixed Residential and Industrial Areas*, British Standards Institute.

39 Christie, D. (1982) Limitations of BS 4142, *Proc. IOA*, 8 February.

40 Bramer, T.P.C. and Eade, P.W. (1982) The Use and Abuse of BS 4142, *Proc. IOA*, 8 February.

41 Greater London Council (1985), *Code of Practice for Pop Concerts*, GLC.

42 Griffiths, J.E.T. (1985) Noise Control Techniques and Guidelines for Open-Air Pop Concerts, *Proc. IOA*, Vol. 7, Part 3.

43   Griffiths, J.E.T. and Kamath, S.S. (1987) Revised Environmental Noise Guidelines for Pop Concerts, *Proc. IOA* – Reproduced Sound 3.

44   Griffiths, J.E.T., Turner, S.W. and Wallis, A.D. (1986) A Noise Control Procedure for Open-Air Pop Concerts, *Proc. IOA*, Vol. 8, Part 4.

45   Greater London Council (1979) *Disco Rules OK? Code of Practice for Discotheques*, GLC.

46   Noise Advisory Council (J. Bickerdike) (1986) *Draft Code of Practice on Sound Levels in Discotheques*, HMSO, London.

47   Bickerdike, J. and Gregory, A. (1979) *An Evaluation of Hearing Damage Risk to Attenders at Discotheques*, Department of the Environment Contract Report.

48   Stirling, J.R. and Craik, R.J.M. (1986) Amplified Music as a Noise Nuisance, *Proc. IOA*, Vol. 8, Part 4.

49   *Code of Practice on Noise from Model Aircraft* (1982) HMSO, London.

50   *Code of Practice on Noise from Audible Intruder Alarms* (1982) HMSO, London.

51   *Code of Practice on Noise from Ice-Cream Van Chimes, etc.* (1982) HMSO, London.

52   Watson, A.E. (1986) The Control of Motor Sport Noise, *Proc. IOA*, Vol. 8, Part 4.

Chapter 4

SOUND INSULATION BETWEEN DWELLINGS

Stephen Rintoul, London Borough of Lewisham

Inadequate sound insulation between dwellings is one of today's main problems in the built environment. It is probably the second most common source of complaint and dissatisfaction (transportation noise being the most common) and can give rise to severe noise problems. In London alone it is estimated that in excess of 50 000 dwellings have poor or unacceptable levels of sound insulation. Nationally, similar problems are experienced in almost every large city or town. Many problems are a legacy from earlier times, times when there was little or no grasp of the likely outcome of constructing dwellings without proper regard to sound insulation provision. However, dwellings are still being created today which do not have acceptable insulation. Developers, local authorities and even housing associations still insist on making short-term financial gains by dispensing with the provision of sound insulation in dwellings for which they are responsible; leaving the occupants of such dwellings to live with the resulting noise problems.

Many people have found that this short-term financial advantage is soon lost when remedial works must be adopted. It makes practical, financial, legal and social sense to get the insulation right to begin with. The statement 'prevention is better than cure' is certainly applicable to sound insulation. It is also true that prevention is cheaper than cure, in most cases significantly so; prevention is always *easier* than cure.

This chapter deals with the provision of sound insulation between dwellings; and between dwellings and common parts. The basic mechanisms of sound transmission are described and the various means of controlling both airborne and structureborne noise are discussed. Consideration is given to the insulation performance of individual building elements (walls, floors, and so on) and to the provision of insulation by proper layout design planning.

Transmission paths and mechanics

To understand the reasons for poor insulation occurring between dwellings and the differing effectiveness of various types of construction it is first necessary to understand the mechanisms of sound transmission in buildings. Such knowledge should enable potential problems to be predicted and preventive measures to be specified at the design stage, and facilitate the identification of the various transmission paths and specification of appropriate remedial works for an existing problem.

In either situation, without a thorough appreciation of the mechanisms of sound transmission, the adoption of insulation techniques which are both effective and economical is unlikely.

The transmission of sound through a building has several components (Figure 4.1).

1   Source.
2   Source room (or space).
3   Transmission pathway(s).
4   Receiving room (or space).
5   Receiver.

The first stage in dealing with any noise problem is to identify potential sources, receivers and the particular mechanisms by which acoustic energy will enter the building structure. It will then be necessary to categorize the transmission pathways and their relative importance (airborne and structureborne noise, direct and flanking transmission, structural and non-structural flanking transmission). Only after identifying the number, nature and type of pathways is it usually possible to assess their relative importance. Finally, the acceptability of the transmitted sound level in the receiving room must be assessed.

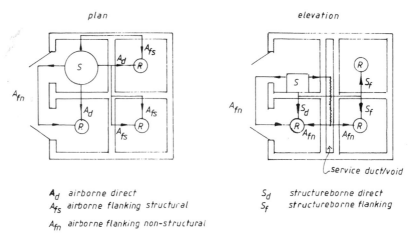

A_d airborne direct
A_{fs} airborne flanking structural
A_{fn} airborne flanking non-structural

S_d structureborne direct
S_f structureborne flanking

Figure 4.1   *Transmission paths*

Airborne and structureborne noise

The main distinction between airborne and structureborne noise is the mechanism by which the acoustic (or vibrational) energy enters the building structure. For a given source both airborne and structureborne noise are likely to be generated. Their relative importance will depend, initially, on the degree of direct coupling between the source and the surrounding structure. The subsequent transmission pathways will dictate how and where the acoustic energy is re-radiated.

Airborne noise is radiated into the source room and, depending on the nature of the boundary surfaces (walls, floor, ceiling), a proportion of the incident sound energy enters the structure and is transmitted through the building, re-radiating as noise into receiving rooms and spaces.

Structureborne noise includes all instances where the acoustic energy, in the form of vibrational energy, enters the room boundaries without an intervening air path. This includes situations where the coupling is a result of impacts on the structure. The majority of the vibrational energy will be transmitted directly into the structure. The transmission of the sound energy thereafter to the receiving room is identical to that for airborne noise.

Structureborne noise is sometimes described as 'footfall' or 'impact'. These terms come from the early studies of sound transmission where the principal source of structureborne noise was the impact of footfalls. This is also reflected in the alternative names for standard source machines, such as "footfall' or 'tapping machines', and the terminology used in some codes and standards of airbone and impact noise.

The terms airborne and structureborne describe the mechanism by which the acoustic energy enters the building structure, not how it is then transmitted, or usually how it is re-radiated. Structureborne is also sometimes used to describe transmission paths. Nearly all transmission paths at some point are structure borne; the use of the term in this context can lead to confusion and should therefore be avoided.

Resistance to airborne acoustic energy entering a structure is generally controlled by the mass of the constructional element (wall or floor) separating the source and receiving rooms. The more massive the element the more difficult it is for the airborne acoustic energy to force that construction into vibration and less acoustic energy will be transmitted. Resistance to airborne noise may therefore be improved by constructing the walls and floors of the building from heavy materials.

With structureborne noise the source is coupled to the building structure, and simply increasing the mass of the building elements may not significantly increase their resistance to structureborne noise. It is necessary to decouple, or isolate, the source of the noise from the structure, ensuring that there are no rigid connections between the source and the structure and, in the special case of impact noise, means avoiding rigid connections between the surface suffering the impact and the rest of the building structure.

Direct and flanking transmission

Once the acoustic energy enters the structure it will be transmitted through the various structural components of the building and can be re-radiated into any other room or space contained within the building.

Where two rooms or spaces have a common boundary surface acoustic energy will be transmitted directly through the common boundary. This is known as *direct transmission*. A proportion of the acoustic energy will also be transmitted via other paths and enter the second room through the other room surfaces. This phenomenon is known as *flanking tranmission* (see Figure 4.1).

Depending on the intensity of the acoustic energy received via flanking transmissions the insulation of any common structure can appear to be much lower than expected from its construction. Where flanking transmission is appreciable, or where there are main flanking transmission paths, problems may arise in distant rooms while there is no apparent problem in rooms adjoining the source room.

Most flanking paths are usually readily identifiable at the design stage. However, the relative importance of the various flanking paths, compared to any direct path, is notoriously difficult to predict, and the existence of flanking transmission will often be the limiting factor on the effectiveness of the sound insulation between rooms.

Structural and non-structural flanking transmission

Flanking transmission via structural elements of the building can be reduced by using the same techniques as for controlling any direct transmissions. Introducing discontinuities and increasing the mass of the components can achieve significant improvements. Flanking transmissions via non-structural paths, such as service ducts, electric conduit and even external openings, are usually more difficult to identify and control. For example, windows in adjoining rooms located close together on an external facade. The wall separating the two rooms may have an acceptable acoustic performance but the windows, especially when open, will provide a flanking path for sound energy. Even when a wall or floor is constructed in a satisfactory manner, service runs, ducts, conduits, and pipes may pass through the construction and give rise to acoustic weak points or flanking paths. All the potential sources of flanking transmission should be identified at the design stage and treated, or the layout redesigned so as to eliminate the pathways.

For remedial works a more limited range of options is available. The design and layout of both the building and the various services will already be fixed and it will not usually be possible significantly to alter either. Rather the acoustician must look to the individual noise sources and transmission paths. It may be necessary to rectify installation faults or design retro-fit treatments. It may be necessary to carry out several stages of treatment, starting with the most obvious pathways first, assessing the effectiveness of the treatment and then extending the treatment to other pathways as may be required.

Assessing sound insulation

Criteria and standards

Criteria for sound insulation in buildings are based on what the occupants of the building find subjectively acceptable. For dwellings this will generally be protection from disturbance by the activities of neighbours and the preservation of privacy. Normal domestic activities will give rise only to occasional minor inconvenience. More serious noise problems should arise only as a result of anti-social behaviour by neighbours or a malfunction of mechanical services and appliances. Criteria are subjective requirements determined by subjective assessments.

Standards attempt to set appropriate objective numerical values to the levels of sound insulation which match the subjective criteria. The standard may be in the form of an exact level of sound insulation, or it may define a class, grade or range of levels within which the same subjective response is expected. The sound insulation performance of a given structure is obtained by objective measurements and rating procedures which usually yield a single numerical figure. This performance level (or rating) of the sound insulation is then compared against the standard to determine likely subjective acceptability.

Standards are usually defined following social surveys of occupants of buildings where the insulation levels are known.[1] Such a procedure introduces a statistical element to the standard. Most standards are set at a level which will be found acceptable by a certain percentage of people. In the case of the old British grade system,[2] Grade 1 insulation would be expected to satisfy 73 per cent, and Grade 2, 50 per cent of occupants. Standards may be required by legal controls, recommended by codes of practice, or set from experience of similar situations.

An agreed measurement and rating procedure enables the acoustic performance of different constructions and alterations to be quantified. This information can be used to specify acceptable forms of construction, to quantify deficiencies in performance and to determine the acoustic requirements of any remedial works. Insulation techniques can be developed and new forms of construction can be tested prior to adoption.

The methods of measuring sound insulation distinguish between airborne and structureborne sound, and standardized measurement techniques are laid down, or recommended, by various national and international organizations.[3,4] Before embarking on any measurement of sound insulation one should be fully conversant with the requirements and limitations of the procedure used and confirm that the procedure is the most appropriate as the results from one procedure may not be readily convertible to those obtained from another.

For airborne sound the basic parameter of interest is the *sound transmission loss* (frequently abbreviated to transmission loss or TL) of the building component under test. Transmission loss is frequency dependent and therefore measured in various predetermined frequency bands (octave or one-third octave). To allow for the variability of acoustic conditions in the source and receiving rooms measurements are corrected for reverberation times in these rooms to give a 'normalized' or 'standardized' transmission loss.

For structureborne sound it is almost impossible to make a direct measurement of the transmission loss of a building element. It is therefore usual practice to measure a secondary parameter which is more readily obtainable; the level of sound in the receiving room, or space, generated by a standard tapping machine in the source room. Such levels are referred to as *impact transmission levels* or *impact sound pressure levels*.

The application of a tapping machine as a standard source representing all forms of structureborne sound is open to question and its application is restricted to measurements of impact sound insulation of floors. To study the effects of other sources of structureborne noise, or the impact insulation of a building element other than floors, non-standard techniques may have to be adopted.

As with airborne sound insulation, impact transmission levels are normalized for receiving room acoustic conditions, and measurements are made in predetermined frequency bands (octave or one-third octave).

Rating sound insulation

Rating procedures are the means of relating objective performance to subjective response. All standardized measurement techniques have a complementary rating procedure or method. The rating procedures require the test results to be compared against a reference performance curve. The rating system may give a single figure, usually known as an *index*, or ascribe the results to a band of performance, usually known as a *grade* or *class*.

The reference performance is usually derived from the insulation of a standard construction, for example 9-inch solid brick wall plastered on both sides for walls, and 100mm solid concrete slab with plastered soffit and resilient floor finish for floors. It should be noted that different rating systems sometimes use different references.

Separate rating procedures are laid down for airborne and impact sound insulation and separate results should always be quoted. Different constructions which have the same insulation rating should provide subjectively similar levels of insulation. Ratings therefore make it possible to rank-order the insulation performance of various constructions and structures, and facilitate the setting of objective standards.

Airborne insulation ratings

The British grade system[2] has been officially superseded by an index system (see below) and is now little used, but will be encountered in older reference material. The grade system defines three grades of insulation performance for floors (Grade 1, Grade 2, Worse than Grade 2) and four grades of insulation performance for walls (House Party Wall Grade (HPWG), Grade 1, Grade 2, Worse than Grade 2).

In addition to ascribing a grade to the insulation performance this rating system can also define a single-figure performance; the *aggregate adverse deviation* or AAD. The AAD is the arithmetic total of the individual adverse deviations, in the predetermined frequency bands, from the grade reference and should be quoted, for example, as AAD = 58 dB with respect to Grade 1, AAD = 62 dB with respect to HPWG, and so on. The lower the AAD the better the insulation.

The index system[5] superseded the grade system in 1980 and was revised in 1984. This rates the insulation performance of floors, walls, and building elements in terms of *airborne sound insulation indices*. These are based on measurements of the *sound reduction index* (R) for building elements or the *level difference* (D) between rooms. This rating procedure yields a single-figure performance index, which is qualified to indicate the measurement conditions. The value of the index increases as the insulation improves. The insulation provided by a structure can also be rated in terms of the *margin* (M), M = 0 is the rating of the reference curve, positive values of M indicate performance superior to the reference, negative values of M indicate performance inferior to the reference.

The current international (ISO) rating system[6] is for all practical purposes identical to the equivalent British Standard. The previous system rated constructions in terms of the *airborne insulation index* (I_a), which is a single-figure performance index, and is closely related to most of the older rating systems. Although not current it may be encountered in older reference material.

The US system[7] rates constructions in terms of the *sound transmission class* (STC). It uses a wider frequency range than either the equivalent British or international systems but usually yields a rating numerically equal to the airborne insulation index. The numerical value of the STC increases as the insulation improves.

Impact insulation ratings

The British grade system[2] defines three grades of impact sound transmission of floors (Grade 1, Grade 2, and Worse than Grade 2). It is also possible to rate performance in terms of AAD.

The current index system,[8] which superseded the grade system in 1980 and was revised in 1984, rates the performance of floors in terms of the *impact sound insulation index* and is based on measurements of the *impact sound pressure level* (*L*). This rating procedure yields a single-figure performance index which is qualified to indicate the measurement conditions. The index is based on the level of transmitted sound, therefore the value of the index decreases as the insulation improves. It is also possible to rate insulation in terms of the margin (*M*), as with airborne insulation.

The current international (ISO) rating system[9] is for all practical purposes identical to the equivalent British Standard.

The US impact isolation class (IIC)[10] rates the impact insulation in terms of a single-figure rating. The higher the numerical value of the IIC the greater the impact insulation of the construction. The IIC is related to the companion airborne rating system, STC, in that numerical equivalence in the ratings indicates approximate subjective equivalence in airborne and impact insulation.

The impact noise rating (INR) is an older rating system[11] developed for the Federal Housing Administration, and rates floors in terms of the INR, a single-figure rating which can take positive or negative values. The rating INR = 0 is given to a floor-ceiling construction which provides satisfactory impact insulation.

Equivalence between the rating systems

It is sometimes necessary to convert ratings from one system into another. If the raw data from the measurements of the transmission loss are available the conversion should be relatively simple.

However, if only the ratings themselves are available the task is much more difficult and only approximate comparisons may be possible. Where there is a marked degree of similarly between the system, for example British and international indices, there may be a simple numerical equivalence. Where there are fundamental differences between the procedures, for example IIC and INR, 'rule of thumb' may have to suffice.

Table 4.1 provides conversion factors for selected ratings.

Standards for sound insulation

The acoustician must know where standards are to be found and which are appropriate to any given situation. When considering objective standards for sound insulation the various rating systems can be divided into those which have an integral standard, and those systems which rate the insulation without an explicit reference to a standard.

In the case of the British grade system any structure which attains Grade 1 insulation will be considered by most people to be satisfactory; Grade 2, Worse than Grade 2, and 8 dB worse than Grade 2, can be used to describe a performance inferior to Grade 1 and should not be considered as acceptable standards of insulation. If the AAD alternative is used the AADs should always be quoted with respect to the reference curve, that is, the AAD indicates the degree of inferiority of the insulation. However, it should be noted that

Table 4.1

Conversion factors for selected insulation ratings

From	To	Conversion Factor or Expression
Airborne ratings		
British grade	British index	Grade 1 $\simeq D_{nT,w}$ 53 (walls) $\simeq D_{nT,w}$ 52 (floors)
British index	International index (revised)	Identical
STC	International index (original)	STC $\simeq I_a$
International index (original)	International index (revised)	$I_a \simeq D_{nT,w}$ $= R'_w$
Impact ratings		
British grade	British index	Grade 1 $\simeq L'_{nT,w}$ 61
British index	International index (revised)	Identical
IIC	International index (original)	IIC $\simeq 115 - I_1$
IIC	INR	IIC $\simeq$ INR + 51
International index (original)	International index (revised)	$I_1 \simeq L_{nT,w}$ $= L'_{n,w}$

the performance of a building element can still be classed as Grade 1 with an AAD of up to 23 dB.

The American INR can be similarly used. The INR = 0 rating is assigned to a construction of minimum acceptability. Constructions which have a negative INR rating can be considered as inferior and unacceptable, whilst those with a positive INR rating will be superior and acceptable. The magnitude of the INR rating will indicate the degree of inferiority or superiority.

For other rating systems there is usually no 'inbuilt' standard, the performance of the structure will be rated by a simple number but its acceptability will not be explicit. The British and international index systems can also rate insulation performance in terms of the margin. This concept of rating is similar to INR; and it could be argued that the margin represents an integral standard although it is not referred to as such.

In most countries there exist well-established legal controls governing the construction and design of buildings. Such legislative controls usually include standards for sound insulation which must be achieved in all new dwellings. The requirements of the legislation may be expressed in general terms or they may set an exact performance standard. For example, the Building Regulations,[12] which apply to new buildings in England and Wales, require that party structures provide reasonable resistance to the transmission of both airborne and impact sound; a general requirement. Constructions are described which meet this general requirement and objective standards of acceptability are also given for other constructions.[13] The objective standards are sound transmission (insulation) performance values and are in the form of ratings as derived by the British index system.

Standards may also be required by local controls. These may set a higher objective standard than any national control (rare), or may set standards for situations not covered by any national control (more likely). Planning controls are often used in this way;[14] local authorities setting their own standards for sound insulation either for a class or type of building, or specific to an individual project.

For remedial works to existing buildings the legal situation is not so well defined but the standards set for new buildings should also be seen as reasonable design goals for remedial works. However, depending on the exact wording and scope of the building control legislation, the standards may not be applicable or enforceable in existing buildings. Where no such requirement exists improvements to the sound insulation are usually controlled by the concept of legal nuisance, which will also control those instances of inadequate insulation, in existing buildings, that can be remedied by recourse to legal action.

One of the important roles of the acoustician will be to ascertain whether any specific legal standards for sound insulation apply to the particular project under consideration. Whatever the situation any standards required by legislation should be seen as minimum standards of acceptable insulation performance.

Standards required by client groups

Sometimes the requirements of the client,[15] or user groups of the building, will indicate that an insulation performance greater than that dictated by legal controls is necessary. In this situation it will be necessary to translate those requirements into design goals for the project and to derive specific standards. Once a standard has been set it will be a design goal for the project and should be included in all design briefs and other related documentation.

Standards recommended by expert bodies

A wealth of reference material exists in the form of recommendations for insulation requirements published by expert, trade and research bodies. This information may include recommended standards of insulation for different situations; it may also include advice on how to achieve given levels of insulation. Such information should always be regarded as purely advisory and its applicability should, whenever possible, be verified prior to adoption.

Guide values for sound insulation standards

From the above it can be seen that the setting of standards for sound insulation is not straightforward. Taking the British index system as the method of rating and conventional residential occupation as the building usage, the guide values in Table 4.2 are suggested.

Table 4.2
Guide values for insulation between dwellings

Airborne sound $(D_{nT,w})$		Impact sound $(L'_{nT,w})$	Subjective description*	Acceptability† (%)
Walls	Floors			
			'Very good'	
57	56	57		90
			'Good'	
53	52	61		75
			'Reasonable'	
49	48	65		50
			'Poor'	
44	44	69		20
			'Bad'	
40	40	73		0
			'Intolerable'	

*Typical subjective response for the 'average' occupant.
†Extrapolated from limited data; use with caution.

Insulation by design

This section deals with the provision of sound insulation between dwellings at the design stage. For rehabilitation projects the proposals assume that all dwellings are unoccupied and the sound insulation provisions are to be incorporated as part of a comprehensive scheme of improvement works to an existing building structure.

Design principles

To ensure successful design of adequate sound insulation one must pay due regard, in a logical and consistent fashion, to each of four elements:

1 Project definition.
2 Layout design.
3 Detail design.
4 Design feedback.

Project definition

In the case of large-scale projects, before commencing any detailed work there must be a clear understanding of the intended use of the buildings and the objectives of the project. Design liaison and co-ordination between the various disciplines should be established to help prevent problems by ensuring that no part of the project is designed in isolation.

For such works the client will usually instruct a consultant architect, who will develop a design brief based on the client's requirements. The consultant architect will usually assume general supervisory responsibilities for the project and perform the central co-ordinating role for the various engineering disciplines.

The design brief should contain information on such things as site location, site infrastructure, size and type of buildings, usage of buildings, general constructional requirements, project time scales and any particular requirements of the client. Wherever possible the design brief should also include relevant performance targets.

It is the role of the consultant to recommend appropriate goals and confirm their acceptability with the architect and make recommendations regarding protection of dwellings from external environmental noise sources, including the layout of the site, vehicular access and building orientation.

When dealing with smaller works (single dwellings, in-fill developments and so on) the same considerations apply. In practice, however, the process is likely to be more informal and the architect may assume all design functions. An acoustician may be consulted for general advice or to assist in the detailed design but, in whatever capacity, should start by obtaining agreement on appropriate design targets for noise criteria and sound insulation.

Layout design

Having agreed appropriate insulation targets the next stage is to consider the internal layout of the proposed buildings. The different areas within the building envelope should be categorized according to proposed usage; residential accommodation, common parts, service areas, storage and so forth. Residential accommodation should be further categorized to include identification of room types; bedrooms, living rooms, bathrooms, kitchens and so on.

Party structures separate dwellings from one another, or dwellings from common areas and other spaces. Wherever possible party structures should be located so as to coincide with the main structural components of the building. This will ensure that party structures benefit from the inherently superior sound insulation of load-bearing elements. Where this coincidence does not occur the project architect should be consulted to determine if the location of the party structure can be changed. If this is not possible the component should be identified for detailed design consideration at a later stage.

The lateral layout of rooms and spaces should be used to optimize the effectiveness of the sound insulation. Proper lateral zoning will prevent room types with dissimilar classifications being adjacent. Consider Figure 4.2, where the layout of the flat has been designed so that the living room (noise sensitive) is protected from the common parts (noise generating) by the kitchen (noise tolerant). Bedroom 1 (noise sensitive) is protected from the common parts by the bathroom and cupboard and Bedroom 2 is protected by the rest of the flat.

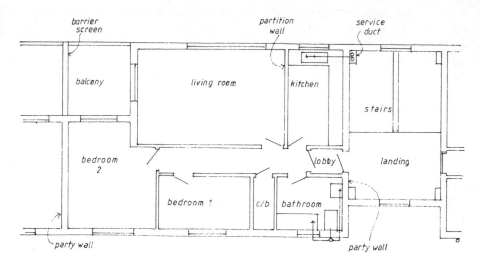

Figure 4.2 *Layout design and zoning*

As with lateral zoning, stacking of room types can also be used to minimize the likelihood of noise problems. The most severe noise problems usually result from improperly stacked rooms, such as living rooms above bedrooms, or bathrooms above living rooms. For flats the simplest approach is to have a standard layout for each floor. This will give a design where bedrooms are above bedrooms, living rooms above living rooms and so on, thus avoiding the worst problems. When considering individual dwellings which are on several levels, such as maisonettes or scissors flats, the stacking arrangement should ensure that different room types in different dwellings are not adjacent. This will necessitate a different floor layout for each level, although the layout of pairs of dwellings may be repeated in the stack. It should be noted that proper stacking of dwellings, particularly flats, will also simplify service design and layout.

Common areas are those spaces and areas in a building to which all occupants, and perhaps the public, have access. Staircases, entrance lobbies, access landings and corridors are all common areas. Such areas typically carry high pedestrian traffic levels and are therefore prone to structureborne noise, particularly as a result of footfalls. Airborne noise may be a problem in some situations but usually much less than any associated structureborne noise.

The layout of common parts needs some special consideration in the design when considering zoning and stacking. No residential space should be located below common parts or circulation areas. Where this cannot be avoided only those rooms which may be classified as noise tolerant, for example bathrooms, kitchens or private entrance lobbies, should be accepted below any common parts. In designs which have dwellings in basements, part of that dwelling will inevitably be below the ground floor common access and circulation areas. These residential spaces should be used only for noise-tolerant purposes. This may result in a different floor layout design for such dwellings; with, say, reduced bedroom or living room size; but such potential shortcomings of the dwellings are more than compensated by the avoidance of at least one source of serious noise problems.

Acceptable sound insulation should be achieved if the walls and floors forming the common area are treated as party structures. This means that lightweight constructions should be avoided for both walls and floors and a resilient layer or finish incorporated into the floor design. Stairs in communal use must be treated in the same way as the floors by introducing a resilient layer. In addition attention must be given to all surface finishes (walls, floors, ceilings). Careful selection of finishes can increase acoustic absorption and this will reduce any build-up of reverberant sound.

Within the common areas of tower or apartment blocks it is usual to find passenger lifts. Any mechanical plant located in the common parts should be positioned as far away as is practical from residential spaces. The layout should also interpose as many structures and noise-tolerant spaces as possible between the mechanical plant and noise-sensitive rooms. Noise becomes a problem when the machinery is particularly noisy or when the enclosing structures do not provide adequate insulation. Some noise will always be generated and the designer must ensure that the enclosing structure provides adequate insulation. This will require that the lift shaft and motor room are of a heavy construction and that any connections to the adjoining structure are kept to a minimum and provided with adequate isolation breaks.

If the design is properly stacked the layout of the principal services will be simplified. Where bathrooms are above bathrooms, and kitchens above kitchens, the main services (gas, electricity, water, drainage) can run vertically and enter each dwelling by a short horizontal service branch. To minimize flanking transmission the main vertical run should be located outside the dwellings, either on the external facade of the building, or via the common parts. The main runs can be grouped together and located in a common service duct, which should avoid noise-sensitive areas. Where any service run passes through a building element that point must be identified for detailed design consideration at the appropriate stage.

Once the service has entered an individual dwelling the service layout should ensure that branch services should pass through noise-tolerant spaces only, and should be protected by ducting. This is reasonably easy to ensure for water or drainage services as their provision is not usually required in living rooms or bedrooms. Their location can be restricted to bathrooms or kitchens and if the building layout includes lateral zoning the services will not need to run through other rooms. The provision of these services for secondary use in, say, a bedroom (en-suite showers or baths) should then be designed so that no party element is affected, and the branch run is located within a noise-tolerant area. Depending on the layout of the dwelling it may be better in these cases to run separate main and branch services for secondary uses.

Electricity, and sometimes gas, services will be provided in all rooms and such circuits should be run via noise-tolerant spaces and not through or within party structures. In practice this is difficult to achieve and the services must be run in suitable ducts and conduits to preserve insulation. The designer must also specify that there should be no back-to-back installation of fittings on party walls.

Detail design

Having determined a suitable layout, consideration should be given to the detailed design of the various elements which make up the building structure. The form of construction of each element must ensure that adequate insulation is provided, and that no weak points or flanking paths will be introduced. Areas where insulation is inadequate will need to be identified and alternative forms of construction or methods of improving the performance of the proposed structure recommended.

Floors   The structural requirements of a floor will be determined by its dimensions and the imposed loads to which it may be subjected. The floor must also provide adequate sound insulation, or isolation, between the spaces it separates. The acoustic requirements of the floor construction will largely be dictated by the uses those spaces are intended to perform and the type of occupancy.

To the acoustician, party rather than partition floors are of primary concern. Such floors require a construction that provides insulation for both airborne and structureborne noise to an appropriate standard. The available forms of floor construction dictated solely by structural considerations will have varying acoustic characteristics and the final selection must be based on their acoustic performance.

Party floors will invariably require additional treatment. Floors must provide protection from not only airborne and structureborne sound but also the effects of impact noise from footfalls. If a floor is of solid concrete slab construction the mass of the floor should be sufficient to control direct airborne noise transmission. In such cases it should be necessary only to introduce a resilient layer within the design to increase the impact noise insulation to the required standard. The exact location of the resilient layer is not critical provided it isolates the main structural components from the effects of impacts on the floor surface. A floor can be provided with a resilient surface finish, or a resilient layer between the structural slab and any floor screed.

For designs based on the latter it is essential that the resilient layer also isolates the impact surface from the perimeter structures so as to minimize flanking transmissions. The materials used for resilient layers should not deform under load to the extent that bridging occurs, nor should the resilient property of the material deteriorate with time.

Concrete floors may be other than solid slab construction; reinforced concrete planks, hollow pot, precast hollow, channel slab and channel beam may all be encountered. For the heavier forms of construction it should be sufficient to add a resilient layer as for solid slabs. The lighter constructions will require additional mass, loadings permitting, which may be provided by appropriate floor screeds or ceiling finishes. If mass cannot be added the designer should incorporate a floating floor finish which will improve both the airborne and structureborne (impact) insulation performance of the lighter floors.

Traditional suspended timber floors invariably have poor insulation for both airborne and structureborne (impact) noise. Although the design can usually incorporate some additional mass the increase will not usually be sufficient to control airborne noise adequately. In addition to any mass increase it will also be necessary to introduce structural discontinuities and resilient layers – floating floors, pugging, resilient material, secondary ceilings – either on their own or in combination to achieve the desired level of insulation. With lightweight constructions it is essential that there are no air paths around or through the structure – appropriate edge and floor covering detail must be specified.

If, unavoidably, one of the room uses is particularly noise sensitive or noise generating (that is, there is potentially a severe noise problem), the design of separating floors should be dictated primarily by the acoustic requirements of the situation. It will be necessary in such situations to design a floor of high insulation which of itself may give rise to increased structural requirements.

Flanking transmission is always a potential problem with any floor. In addition to any acoustic weak points, such as service runs, the perimeter construction is critical if flanking transmission via walls is to be avoided. The design of floor-wall junctions is an integral part of the acoustic design of the floor. Both the structural joint between the floor and wall and any insulation treatment incorporated in the design need to be properly detailed if flanking effects are to be minimized.

The structural requirements of the junction will often limit what is achievable but alternatives should always be considered before a final design decision is made. At all stages of the actual construction works adequate site supervision is necessary to prevent poor workmanship or unauthorized variations degrading the expected insulation.

Walls Party walls usually perform a structural support function and most will be of load-bearing construction. Such party walls, if constructed to current standards, will also provide acceptable airborne noise insulation, with little or no modification. An exception to this is when a very light form of construction is adopted, such as timber frame, where there is insufficient mass to control airborne sound. Here other forms of control such as structural discontinuities must be incorporated to compensate for the low mass.

Such problems as do arise are typically a result of poor workmanship, flanking transmissions, or inappropriate location of mechanical equipment and services. The first can be avoided by adequate site supervision during construction and the others minimized by proper acoustic assessment at the design stage. To remedy these faults in existing structures is more difficult, and may even require substantial reconstruction works if the problem is to be completely solved, though it may be possible to achieve significant improvements by treating each fault or defect on an *ad hoc* basis.

Party walls which are not load bearing are likely to be of lightweight construction. To give adequate party separation the mass of such walls should be increased by either a change in design or simple addition of mass to the structure. Where there is a structural loading limitation the insulation can be improved by increasing the mass of the structure up to the loading limit and adopting a multiple-skin construction incorporating discontinuities and absorbent in-fill to voids.

Partition walls separate rooms in the same occupancy and the acoustic performance criterion is less stringent than for party walls. Partition wall construction is typically of lightweight block or timber studding, finished with a board (dry lining) or wet plaster system. Forms of construction suitable for structural requirements should also provide adequate sound insulation from airborne noise in most situations. Exceptions are where a higher level of insulation is required, for example partitions enclosing a bathroom or water closet. The acoustician may be able to increase the mass of these walls and still keep within design loadings, or it may be necessary to adopt a multi-skin construction.

Services   The detail design of the building services must take account of the three principal ways in which services can cause noise problems. First, the service may be a source of noise (typically plumbing, air conditioning systems, elevator machinery). The control of this type of noise at source is dealt with in Chapter 7. Where it is not possible to eliminate the noise at source the acoustician must specify a construction which will contain the noise. If service ducts or conduits are used they must have an insulation performance sufficient to prevent noise break-out, or a secondary independent duct may need to be provided. It will also be necessary to ensure that there are no rigid connections between the service and the surrounding structure. Mounting brackets and clips should incorporate a resilient layer either between the service and the mount or between the mount and the structure.

Second, service runs may provide flanking transmission paths for noise which by-pass the main structural components. Suitable layout of the service system should reduce the likelihood of this occurring but the problem is almost impossible to avoid completely by layout alone. Additional insulation must be required to prevent noise entering the service system, being transmitted through and breaking out into other rooms. All exposed runs should be ducted, and the void between the service run and the duct should be filled to eliminate air paths. Where a service enters or leaves a duct there must be no air gaps between the service and the duct wall; materials used as fillers in this situation should be able to withstand some differential movement, and non-drying mastic compounds or neoprene gaskets are suitable examples.

Third, where a service run passes through a building element an acoustic weak point may be created. Where the service run is normal to the wall or floor surface, the opening should be just large enough to allow the service to pass through without touching the element and any gap must be sealed with a flexible filler. Secondary ducting should enclose the opening and should also be isolated from the element by a flexible seal. A service that runs through a wall or floor parallel to the surface may not be visible but can still introduce acoustic weak points. Electrical wiring, and occasionally water distribution pipes, may be buried in the building elements. If voids are left surrounding these services the potential insulation of the building element will not be realized.

This problem is made worse by running services back to back through the same building element. If voids are left around the services the effective thickness of the wall may be reduced by a half or even two-thirds. Although the area of the wall affected may be small, the loss of insulation is likely to result in an unacceptable level of insulation between the two rooms separated by the wall.

Secondary ducting of all service runs and pipes is a worthwhile practice provided there is access for maintenance and repair. The design of secondary ducting will depend upon the noise levels expected, and particular attention must be given to the detail of access points if the acoustic function of the duct is to be maintained. Acceptable secondary ducts for most domestic services can be formed by multiple layers of plasterboard fixed to a timber frame, the void between the duct and the service being filled with a mineral quilt (Figure 4.8).

Services such as conventional wiring may run within the structures, and may also have to pass between different dwellings. By careful location of the service run, or acoustic treatment, it is possible to ensure that such runs do not create flanking paths.

Openings Every building has essential openings to allow access, natural lighting, and natural ventilation. These weak points can degrade the insulation of the structure and create flanking paths. Where openings are in partitions, for example room doors, the resultant loss in insulation is not usually significant, but where openings are in party or external walls the loss of insulation is more important.

In the case of windows, a flanking path can be created between two rooms which otherwise have acceptable insulation. The positioning, size and style of fenestration are usually dictated by requirements other than sound insulation. However, for buildings which will be exposed to high levels of external noise, for example railway and road traffic, windows that have good insulation performance are likely to be specified. Such installations will also reduce flanking transmissions between adjacent rooms. Another design feature which can improve insulation from flanking transmission, but which is usually adopted for non-acoustic reasons, is the use of thermal double glazing. Although the acoustic performance of conventional thermal double glazing is not as good as acoustic secondary glazing it can provide a worthwhile increase in acoustic insulation over ordinary single glazing.

Windows provide the lowest insulation when open, and in the worst cases severe disturbance can be caused between rooms within the same or adjoining buildings. This may be counteracted to some extent by adopting a style or pattern of window which provides a partial barrier when open.

The window layout should maximize distances between windows to different dwellings vertically and horizontally. Window design should allow variable areas of opening by subdivision of the total window area into several separate casements. If designed as opposed pairs the casements can when open still provide some insulation, albeit reduced, to noise from adjoining dwellings.

Where private balconies are to be a feature of the building they may be incorporated into the design as acoustic barriers. Such barriers can protect rooms from flanking transmissions from adjoining rooms, and can also protect rooms from some forms of external environmental noise. If a private balcony is to act as a noise barrier it should have end panels to the height of the adjacent window heads; the combined depth of the balcony and height of any *solid* panel to the front of the balcony should be a minimum of two metres.

Most doors in a dwelling are located in internal partition walls. Unless a very high standard of insulation is required, either between dwellings or between rooms in the same dwelling, the construction of such doors is not critical. In fact a lightweight door, complete with a lightweight latch system, may reduce the incidence of annoying door slam. Where internal doors are of heavy construction, for decoration or fire protection, the door frame or lining should incorporate a resilient stop and the striker plate should be isolated from the frame.

In some situations mechanical door closers, with a two-stage closing action, have been employed to reduce door slam. Suitable closers shut the door rapidly until just before the latch engages the striker plate, a damping mechanism then slows the door so that the final seating of the door takes place with the minimum force. While it would be difficult to isolate the door frame from the surrounding wall it may be worth considering isolating the frame from the floor construction by interposing resilient pads between frame and floor.

Entrance doors to dwellings, whether located in external walls or leading off common areas, must provide security and fire protection. Consequently entrance doors should be of heavy construction, well fitting in the door frame, and the frame should be securely mounted in the surrounding wall.

Where common or circulation areas are expected to carry high volumes of pedestrian traffic and activity the individual entrance doorways should be designed to a higher level of insulation than can be achieved with conventional heavy doors. The best practical solution to this situation is to introduce a second door within the dwelling to form an internal entrance lobby or hallway.

Design feedback

It is essential for a successful design that no element or feature of the building is designed in isolation. The design requirements and recommendations set by the various engineering disciplines must be brought together and developed into a single project design. From the earliest design stage there should be feedback from each engineering discipline (structural, mechanical, services, acoustic, and so on) on their specific design requirements and the implications of these requirements for each of the other disciplines. All design decisions, no matter how minor, should be communicated to all members of the design team and there should be design review meetings at each main stage of the design exercise. It is time consuming and costly to have a structural engineer spend time on the detailed design of a floor construction only to have the acoustic consultant recommend at a later date that a 50 per cent mass increase (or some other equally fundamental change) must be achieved.

In many situations design decisions will be compromises, for example the design of entrance doorways or the use of spaces under common paths. In other situations design solutions can be found which meet several apparently unrelated requirements; proper stacking improves sound insulation and simplifies service design; private balconies can provide an attractive feature to the building and improve protection from noise. Design feedback, by formalized liaison and review, will not only ensure a successful design but in most cases will also generate the most cost-effective solution.

Constructional techniques

Each individual design project will have its own constraints and it may not be feasible to adopt a simple 'textbook' approach. One of the skills required is to know which features of a particular form of construction are important and which can be modified without adversely affecting the insulation performance when a one-off construction is necessary.

Chapter 5 also contains a discussion of these effects and possible palliative measures.

Principles

Mass

Any desired level of airborne insulation can be achieved by constructing a component as a single leaf of sufficient mass. In practice this will not always be possible because mass restrictions may be imposed by loading limitations, additional support requirements and space constraints.

Uniformity

The uniformity of construction, and hence distribution of mass, will affect the insulation performance of a building element. In single leaves the objective is to have a uniform distribution so that there are no acoustic weak points. The principle of uniformity also applies to the boundaries of components and any junctions, where all gaps or spaces must be filled to prevent acoustic leakage. The joints between blocks or bricks should have no gaps and ideally the surface of the panel should be coated with an impervious finish prior to any decoration.

For double, or multiple-leaf, construction the simple mass law does not apply. In theory if both leaves were identical, adequately spaced, and completely independent, it might be expected that a numerical doubling of the insulation would result. In practice it is not possible to achieve complete structural independence with conventional construction. However, substantial gains in insulation can be obtained, over that predicted by the mass law, by having a number of homogeneous independent leaves.

Stiffness

For a given construction the insulation is often markedly lower than that predicted by simple theory. The frequencies at which this occurs are determined by the stiffness of the panel and are the result of either resonance or the coincidence effects.

Discontinuities

Discontinuities are particularly important in relation to structureborne noise. Consider a floor-ceiling construction which receives impacts on the floor surface. If the floor and ceiling elements of the construction are structurally independent, direct transmission paths for the impact noise will be reduced and consequently the insulation will be improved. Even where complete independence is not possible significant improvements to the insulation can be obtained by allowing only the minimum bridging across the structure.

Resilience

Another way to control structureborne noise is to use layers of resilient material which deform under mechanical load and thus reduce the energy of vibration. This must not be confused with acoustic absorption – materials which make effective resilient layers may not be good acoustic absorbers, and vice versa.

Many problems have been caused by the mistaken use of one material for both purposes or the wrong application.

To be effective the resilient layer must deform under load and then return to normal when the load is removed; but it must also retain some residual elasticity even when under load, otherwise a transmission bridge will be formed. The resilient properties of the material must not deteriorate with age or worsen with wear. For impact noise on floors the resilient layer may be introduced as the floor finish, or covering, which will greatly reduce the amount of energy entering the floor structure. This may result in a 'spongy' feel to the floor, which could be unacceptable to the occupants. As an alternative the resilient layer may be located between the floor surface and structural components. In this situation the floor surface, which suffers the impact, is isolated from the underlying structure. It is also possible to mount ceilings with resilient fixings and so isolate the ceiling surface from the structure above.

Absorption

Acoustic absorption relies on the porosity of the material and not necessarily on its mass, resilience or flexibility. The role of acoustic absorbers in sound insulation is limited to controlling resonances in multiple-skin constructions.

Types of construction

The remainder of this section considers the construction of typical building elements, describing the important features of the element and how similar features may be incorporated in other forms of construction. Where specific values are given for mass, thickness and so on, they are minimum dimensions for achieving at least a 'reasonable' level of sound insulation, and probably a 'good' level (see Table 4.2), in new-build and large rehabilitation projects.

Single-leaf elements, examples of which are brick, block and masonry walls, and most forms of concrete floors, have sound insulation that relies on its superficial density, stiffness, uniformity and edge conditions (air gaps and bridges to adjoining elements). The performance of the panel in respect of structureborne sound will depend on the effectiveness of any resilient layer(s) in isolating, or decoupling, the main structure of the panel from the noise source.

Multiple-leaf elements are represented by constructions such as brick cavity walls, stud walls and timber joist floors. The sound insulation of multiple leaf panels depends on the properties of each leaf or skin, their mass, uniformity and so on, and the degree of isolation between the leaves. Isolation can be achieved by resilient layers and structural independence of the leaves.

Floors

Concrete floors   All forms of concrete floors behave, acoustically, as single-leaf panels. The main features of such floors are the superficial density, the resilient layer and the floor finish. Where floors are beam or beam and block construction it is assumed that all joints between the components have been filled with an appropriate grout or mortar mix. Figure 4.3 shows diagrammatic representations of concrete floors.

For heavy concrete floors (mass at least 365kg/m²) the basic structure will provide control of airborne sound. The mass of the base structure includes the mass of any ceiling finish or bonded floor screed. Impact sound insulation in floors of this mass can be controlled by the provision of a resilient finish to the floor surface. The resilient layer should have a minimum (uncompressed) thickness of 4.5mm and comply with the characteristics of resilient material

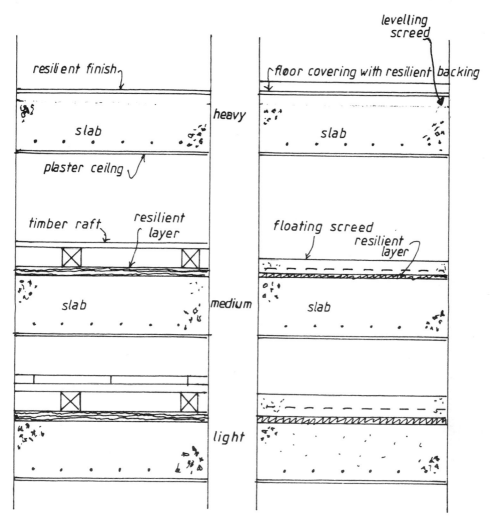

Figure 4.3 *Concrete floors*

as outlined in earlier sections. The layer may form the backing of a floor-covering system; in such cases the mass of the covering system should be omitted from the calculation of the mass of the floor.

Medium concrete floors have a base structure with a superficial density in the range, 300–365kg/m². Such floors will not provide acceptable insulation without the provision of additional mass and isolation in the form of a floating finish. The floating finish can be a timber raft, 18mm tongue and groove timber or timber based board (mass at least 12kg/m²) rafted on 45 × 45mm timber battens, laid without fixing through to the resilient layer. Alternatively, the floating layer can be a reinforced floor screed, thickness 40–65mm (mass at least 75kg/m²). The resilient layer isolating the floating finish from the base structure should be a mineral fibre quilt 13–25mm (density at least 35 kg/m³) covering the whole of the base structure. If a screed is used, sheet closed-cell polyethylene (5mm) or pre-compressed sheet polystyrene (13mm) may be used as alternatives to a mineral quilt.

Light concrete floors have a base structure with a superficial density in the range 220–300kg/m². Such floors may be dealt with in the same general manner as medium floors but because of the lighter base structure a more massive floating finish must be adopted. If a timber floating finish is used the mass of the floating layer should be approximately doubled (to 25kg/m²); this may be achieved by an additional layer of the floor material, or by inclusion of a layer of 19mm plasterboard as a sub-layer to the floor boarding. Screeded floating layers should be increased to a superficial density of at least 130kg/m². Details of resilient layers are as for medium floors.

Very light concrete floors of superficial mass less than 220kg/m² will not provide acceptable sound insulation without fundamental design changes. Such changes usually result in a highly complex design which renders very light concrete floors impractical or uneconomic.

Suspended timber floors Timber floors behave as multiple-leaf panels; for the same sound insulation performance they can be of lighter construction than concrete floors because the materials radiate sound less efficiently and each design described incorporates a greater degree of structural isolation. The important features of timber floors are the mass and location of the individual leaves, the inclusion of resilient layer(s), the provision of a structural discontinuity and absorption in any cavities.

A platform floor (Figure 4.4) should have a base structure which should be 12mm tongue and groove timber, or timber-based boarding (mass at least 8kg/m²), fixed to 50mm-wide timber joists, with a ceiling of 30mm plasterboard in two layers (mass at least 28kg/m²), for example 19mm plank plus 12mm board broken jointed, fixed to the joist soffits. An absorbent blanket, for example fibre glass, with a density at least 12kg/m³, must be laid to a depth of 100mm between the joists. Above the floor base should be a resilient layer formed by a 25mm mineral fibre quilt of density at least 60kg/m³. This layer must not be rigidly fixed to the base structure but may be spot glued to

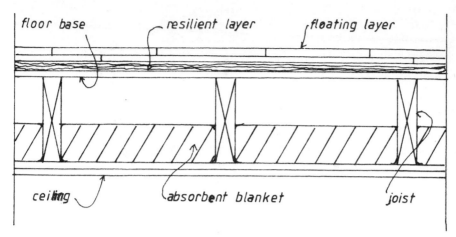

Figure 4.4 *Platform floor*

hold it in position during construction. On top of the resilient layer should be laid a floating layer formed by fixing together two thicknesses of 18mm tongue and groove timber or timber-based board. Alternatively, the floating layer may be formed by 18mm timber board with 19mm plaster plank.

With rafted floor (Figure 4.5) the ceiling, joists and absorbent blanket should be as for platform floors. The floating layer is formed by rafting the floor using 45 × 45mm timber battens isolated from joists by resilient strips of mineral fibre, density at least 70kg/m³. Flooring should be of 18mm tongue and groove timber, or timber-based board, fixed to 19mm plaster plank. The fixings between flooring layers and battens must not penetrate the resilient strips. Resilient strips may be held in place during construction by spot gluing. Alternatively, the absorbent blanket may be replaced by pugging (mass at least 80kg/m²) on polythene sheet, in which case flooring can be a single layer of 18mm tongue and groove timber and the ceiling should be 19mm dense plaster on expanded metal lathing.

Rafted floors with resilient ceilings (Figure 4.5) should have a base structure formed by 50mm-wide timber joists and a floor of 18mm tongue and groove timber, or timber-based boarding, rafted on 45 × 45mm timber battens and isolated from the joists by a resilient layer of 13mm mineral blanket (mass at least 36kg/m³) laid over the joists. The ceiling should be formed from 30mm plasterboard, for example 19mm plank plus 12mm board broken jointed (mass at least 28kg/m²), suspended from joists using resilient bars. Lightweight pugging of inert mineral material (mass at least 15kg/m²) must be laid between joists. Raft fixings and ceiling fixings to resilient bars must not penetrate the joists.

Secondary independent ceilings (Figure 4.6) should have a basic structure of 50mm timber joists supporting flooring of 18mm tongue and groove timber,

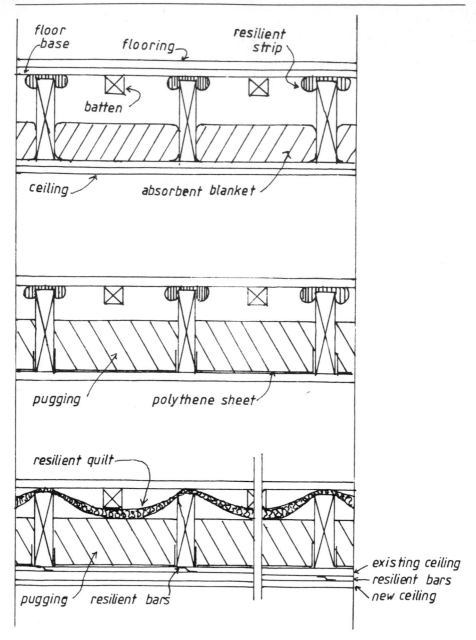

Figure 4.5 *Raft floors*

or timber-based boarding. The intermediate ceiling should be formed from 19mm plasterboard (mass at least 17.5kg/m²) and fixed to joist soffits. The secondary ceiling should be formed from 30mm plasterboard, for example

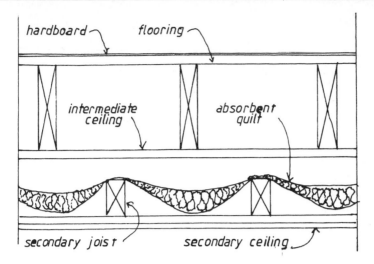

Figure 4.6 *Secondary independent ceiling*

19mm plank plus 12mm board broken joint (mass at least 28kg/m²), fixed to secondary independent joists. Independent joists must not be rigidly connected to the intermediate ceiling or its supporting joists. The gap between the intermediate and secondary ceilings should be at least 150mm. If the gap can be increased to 250mm the mass of the secondary ceiling may be reduced to 22kg/m², for example 2 × 13mm plasterboard. An absorbent blanket, density 12–35kg/m³, should be draped over the secondary joists.

Walls

Solid walls Walls usually need to provide protection only from airborne sound, and in consequence the most important feature of a solid wall is its superficial mass. The different materials that may be used for solid wall construction – brick, block, concrete and so on – have slight variations in performance caused by differences in their mechanical properties. These variations in stiffness, damping and uniformity required variations in the superficial mass of different materials to achieve the same level of insulation. Where the superficial mass of the solid wall is lower than that required, lightweight independent panels may be used to supplement the performance of the wall.

Heavy brick walls should have a mass at least 375kg/m² excluding any wall finish. The wall finish may be a wet plaster system (thickness 13mm) or a similar thickness of plasterboard bonded to the brick surface, applied to both sides of the wall. A bond should be used which includes at least 25 per cent headers and all mortar joints must be complete. For typical bricks (density at least 1610kg/m³) a wall thickness of 225mm will be required to give the desired superficial density.

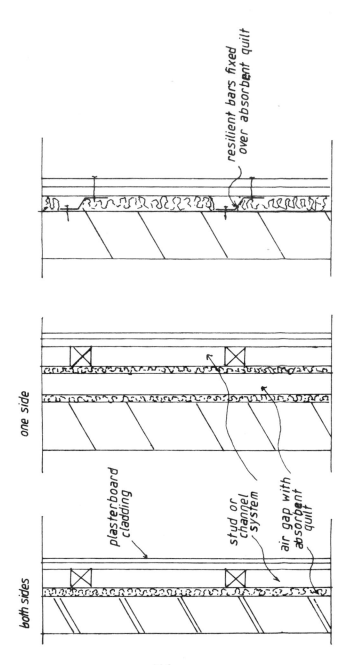

Figure 4.7 Panel systems for walls

Light brick walls have a mass in the range 300–355kg/m², and their performance must be supplemented with lightweight panels. Suitable panels (Figure 4.7) may be formed of two layers of 13mm plasterboard bonded together (other panel systems having similar mass may be used as an alternative), supported on a frame or channel system which is independent of the brickwork, that is, only fixed at floor and ceiling with no contact with the wall. Joints between panel sections must be made airtight either by taping or applying a plaster skim coat. There should be a continuous gap between the brickwork and panels of at least 25mm, which should be filled with a 25mm absorbent curtain (mass at least 12kg/m²); both sides of the wall must be treated.

Heavy concrete walls may be formed by blocks, precast panels or cast-*in-situ* sections. All joints must be grouted and filled, and blocks should be full width, that is, there should be no vertical joints in the plane of the wall. Heavy concrete walls must have a mass of at least 415kg/m² and should be finished on both sides as for heavy brick walls.

Medium concrete blockwork consists of walls formed by concrete blocks (minimum block density of 1500kg/m³) giving a superficial density in the range 300–415kg/m². The usual precautions of filling all joints and using full-width blocks apply. The performance of medium blockwork must be supplemented by the provision of independent lightweight panels to both sides of the wall. Suitable panels and panel systems are as for light brickwork.

For walls formed from lightweight blocks (block density less than 1500kg/m³) the superficial density of the wall must be at least 200kg/m². In the special case of walls formed from autoclaved aerated blocks, the superficial density of the wall can be reduced to 160kg/m². The usual precautions of filling all joints and using full-width blocks apply. The performance of all types of light concrete blockwork walls must be supplemented by lightweight panels to both sides of the wall. Suitable panels and panel systems are as for light brickwork.

Cavity walls Cavity walls are formed by two identical leaves of masonry, typically brickwork or blockwork, separated by an air gap. Because of the air gap it might be expected that cavity walls provide a higher level of sound insulation than solid walls of similar mass and materials, but this is not the case in practice. All cavity walls rely on wall ties between the leaves to provide lateral restraint, and these ties degrade the isolation between the leaves. In addition there will be differences in the stiffness of a solid wall compared with a cavity wall of the same mass and material. To achieve the same insulation performance it is necessary to construct cavity walls to a higher superficial density than their solid counterparts.

For brick and concrete blockwork the superficial density of the wall (both leaves and plaster finish) should be at least 415kg/m². Both sides of the wall should be finished with either a wet plaster system, thickness 13mm, or 13mm plasterboard bonded to the face of the wall. The two leaves of the wall should be separated by an air gap of at least 50mm, which should be free from any material other than the wall ties.

For lightweight concrete blockwork the superficial density of the wall (both leaves and plaster finish) should be at least 250kg/m². Both sides of the wall should be finished as for brickwork cavity construction. The cavity between the leaves should be increased to 75mm.

Timber-frame walls   Timber-frame walls are light-weight constructions which rely on structural isolation between the faces of the wall and absorption in any air spaces to provide sound insulation. Less mass is required than solid or cavity construction, because the materials used radiate sound less efficiently than masonry and there is better structural isolation. Sealing all joints and gaps is critical with this form of construction. Timber-framed party walls should be constructed using the following guidance regardless of whether or not they perform a load-bearing function.

The wall should be formed by two timber frames isolated by an air gap, the sizing of the individual members of the frames being determined by loading considerations. Bridging the air gap by ties must be kept to a minimum consistent with lateral restraint. The spacing of the frames must give at least 200mm between the face claddings. A suitable cladding for each face is 30mm plasterboard (mass at least 28kg/m²) made up from multiple layers bonded together with staggered joints. An absorbent mineral fibre quilt (density at least 12kg/m³) should be installed in the void between the frames. Where the quilt is suspended to form a curtain it should be 25mm thick. If a quilt is fixed to only one frame it should be 50mm thick, if fixed to both frames each quilt should be 25mm.

In load-bearing walls any plywood sheeting or masonry core provided for structural support should not introduce bridges between the frames and should be ignored for sound insulation purposes.

Effects of wall finishes   The constructions described include details of wall finishes. Such systems are typically wet plaster applied direct to brick or block faces, plasterboard bonded direct to wall faces, or plasterboard forming independent panels. These finishes will enhance the insulation of the base structure and form an integral part of the construction.

Most systems of dry lining will not perform acoustically as well as a wet plaster system of the same mass. There will not be sufficient discontinuity to improve isolation, and there may not be sufficient contact to make the dry lining behave as part of the base structure. A finish of plaster direct to the base structure or an independent cladding will provide a superior performance. Some systems of dry lining may even introduce additional resonances, which degrade the insulation of the base structure at specific frequencies.

However, some wall finishes may actually degrade the insulation of the base structure and care must be taken in specifying such details. The fixing of thermal insulation boarding (plasterboard backed with polystyrene) is critical if the base acoustic insulation of the wall is not to be made worse, and it is advisable to install such boarding as independent panels. Adverse effects have also been reported where the insulating material has been fixed to the wall, then plastered over using a wet plaster system, or wood wool slabs have been left as permanent shuttering for concrete.

Service ducts

All surface or exposed service runs, main and branch, should be protected with secondary ducting. The duct provides insulation against airborne sound transmitted via the service run and sound radiated by the service itself. Structureborne noise originating from the service run should be controlled by ensuring adequate isolation at all mountings and at any point where the service passes through a building element.

Suitable service ducts can be constructed using the principles of secondary independent panels (Figure 4.8). For main services the duct cladding should be two layers of plasterboard giving a total thickness of 30mm, for example 19mm plus 13mm board, having a superficial density of at least 28kg/m². The cladding should be supported on a timber frame, which for vertical runs must not be rigidly fixed to the floor or ceiling finishes. Sufficient space must be left between the duct and service run, and the service run and wall, to allow a 25mm absorbent mineral fibre blanket to be wrapped around the service run. Alternatively, the absorbent blanket (mass at least 12kg/m²) may be loosely packed in the void between the service run and duct wall.

For branch services the duct cladding may be reduced to two layers of 9mm plasterboard giving a superficial density of at least 15kg/m². A suitable form of access to ducted services is shown in Figure 4.8.

Junctions

Once a decision has been made on the form of construction for each of the building elements the components must be put together in such a way that their expected insulation performance is achieved. The critical areas of concern are the structural joints and junctions between the building elements. Inadequate attention to detail, or inappropriate junction design, can introduce flanking paths and acoustic weak points, which will seriously degrade the overall sound insulation.

It would be impractical to give a design detail for every possible combination of wall and floor junction, but it is necessary to be aware of the important features of such junctions, and appropriate measures that can be taken to preserve sound insulation. In practice the design of wall-floor junctions will be a compromise between maintaining acoustic separation and structural integrity.

Where a wall meets a floor there must be mechanical joints between the structural components. The joints must be strong enough to support the dead and superimposed loadings and distribute these loadings to the main building

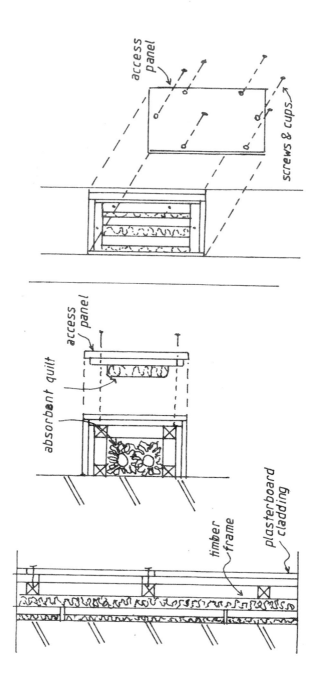

Figure 4.8 *Ducts for service pipes*

structure. The joints will inevitably allow sound to pass from one structural component to another and these transmission paths must be kept to a minimum consistent with structural support. Only those components of the wall or floor which are necessary for structural support should be part of the structural joint. The non-structural components of the wall or floor should not form part of the structural joint. Any junctions involving non-structural components should maintain the independence of those components from adjoining elements. For example, in timber joist floors (see Figure 4.9) the joists should be supported from the perimeter walls by joist hangers rather than building in the joist ends or using timber wallplates.

Some forms of construction and junction design will introduce voids or cavities. Where these cavities are continuous across or through a party element they must be blocked (Figure 4.9) in order to avoid additional transmission paths. Blocking between joist ends in timber frame, or where there are independent wall panels, can eliminate this source of transmission. Note, however, that cavities providing structural independence must be maintained.

All gaps between the components must be sealed to prevent transmission paths. Gaps in structural joints can be sealed with a rigid material, for example, mortar grout or pointing in any gaps left at the joint between a concrete floor and masonry wall. Gaps between other components should be filled using a flexible filler to maintain isolation. In addition, where the junction is between room surfaces such as wall and ceiling, tape or covering may be used to improve the seal (Figure 4.9 and 4.10).

Secondary independent panels are effective because they are isolated from the underlying structural element. At junctions the isolation must be maintained and only the mechanical joints required for support of the panel, usually with adjoining structures, should be allowed.

Floating floors should be isolated from adjoining walls by leaving a small gap between the floating layer and the wall surface. The gap can be filled by turning up the resilient quilt, isolating the floating floor from the floor base at the wall. Alternatively, where the resilient layer is a board material a section of boarding should be cut and used as a resilient pad between the floating layer and wall.

Remedial works

In addition to new-build and large rehabilitation schemes a considerable amount of acoustic design work involves remedial works to existing dwellings where the sound insulation has been found to be inadequate. Although rehabilitation and remedial projects both deal with existing buildings a distinction must be made in the design approach. In the case of the former a scheme of sound insulation will be incorporated in a building programme, which in itself may require fundamental alterations to the structure and layout. It should therefore be possible to include sound insulation provisions in all aspects

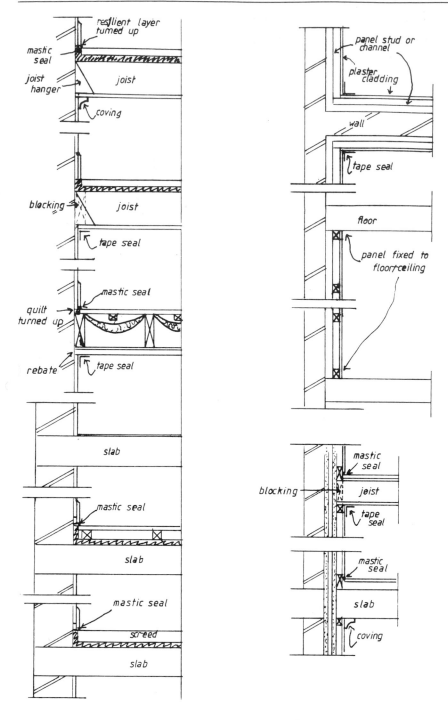

Figure 4.9   *Wall/floor junctions*   **Figure 4.10**   *Panel and duct junctions*

of the design and it will be possible to influence the position and form of construction of each individual component, and the layout of the building. As a consequence, serious defects of the original structure can be rectified and the new elements of the building can be designed to avoid poor insulation.

When considering remedial schemes the available options for improving the sound insulation will be more restricted. There may be loading limitations on the existing structure, there will be little scope for layout alterations and it may not be possible to have access to all parts of the building and its structure. Substantial reconstruction or layout changes are usually out of the question and the work confined to retro-fit treatments to the existing building elements, minor reconstruction, and limited layout alterations. Remedial works present the additional problems of working within fixed practical constraints which limit the available options for improvements to sound insulation.

These constraints will be exacerbated when the remedial works have to be undertaken whilst the dwelling is still occupied. Even when a particular dwelling is vacant for the duration of the works the adjoining dwellings may be occupied and prevent unhindered access to the intervening structures.

Assessing the problem

The first stage must be to assess the adequacy of the existing insulation and the principal transmission paths. This is usually done by a combination of visual inspection and examination of the structure, interviews with the occupants and objective measurements. The original plans of the building and details of any subsequent alterations are also a good source of information (although their accuracy should always be verified on site). On this information the nature and scope of the problem and the necessary remedial works must be assessed. Minimum project design targets of 'reasonable' sound insulation should also be set.

Following this assessment those problems that cannot be solved (that is, attain the design target) without substantial reconstruction or redesign works will be identified. The client must be advised accordingly, the available options explained and a scheme of remedial works developed which provides the optimum increase to the sound insulation, within the existing constraints. Given a building of conventional construction it should be possible to achieve a 'reasonable' level of sound insulation for direct noise; if flanking transmissions are low, and layout design suitable, it should be possible to achieve a general level of 'good' sound insulation.

The client must also accept that once the insulation of the main transmission paths has been improved, the existence of other paths (previously masked by the effects of the main paths) may become evident and this may necessitate a second stage of remedial work. Provision for this eventuality must be made.

Limitations

The limitations imposed on the possible improvement of sound insulation to be attained during remedial works can be broadly grouped into those dictated by the existing layout of the building and those caused by the existing forms of construction.

Design and layout

Rooms and spaces   In most cases the room layout of an existing building will be fixed and there will be little scope for remedying layout design faults. Where there is inappropriate lateral zoning or stacking some interchange of room use may be beneficial, for example the interchange of a bedroom with a living room. There is the disadvantage that interchange may give rise to location of rooms and changes in room sizes being so unacceptable to the occupants that the continued use of the rooms for their new role cannot be guaranteed.

Services   The position of the main services will be fixed and it will not usually be possible to re-route them without substantial reconstruction work. Branch sevices may be more amenable to relocation, but it is usually more cost effective to treat them *in situ*.

Construction

Mass and loadings   The existing components of the building will have structural loading limitations which will be outside the scope of the remedial work. This may limit the mass that can be added to an individual component or supported from an existing structure. The problems of added mass limitations are particularly important when dealing with stud wall or timber floors. These are typically light constructions which do not have good basic insulation and frequently perform the role of party structures.

Thickness and depth   Depending on the position of door heads, window heads and sills there may be restrictions on the available space to install treatments to floors and ceilings. The position of walls relative to door openings and services may limit the depth available to treat the wall. Minimum legal requirements on the height of rooms or the width of corridors may also restrict which treatments can be considered.

Levels   A particular problem encountered when dealing with floors and stairs is that a treatment which raises the floor level will necessitate changes to the doors (that is, they must be trimmed). It may also necessitate refixing bathroom and kitchen fittings to maintain work surfaces and the like at a suitable height.

Where a floor treatment extends to the top of a flight of stairs the topmost riser in the flight will be a different height, and all treads may have to be treated. Differences in floor level between rooms should not be a problem as, for the optimum insulation improvement, all floors should be treated.

Defects   Certain defects in the existing structure, for example voids and transmission bridges, which can give rise to poor insulation may not be identified. Even when their existence is known it will not always be possible to effect a remedy as opening up and exposing the structure may be impractical and may be prevented by the presence of the occupants of the dwelling.

Occupation   From an acoustic point of view, some suitable treatments may have to be rejected if the property is occupied. While the co-operation of the occupants can usually be relied upon as the works will be for their benefit, there is a limit to what most people will accept in the way of inconvenience and disturbance. In some situations it may be better to find alternative accommodation for the occupants for the duration of the works.

Access   It is necessary to maintain access to the existing structure and services for routine maintenance and repair work. Any treatment specified may have to incorporate access points or be designed so that it is not disturbed by maintenance workers gaining access, and is simple to reinstate. Treatments and techniques which do not require specialist skills to install are therefore to be favoured.

Remedial works

General

Many situations can be improved, and some problems even solved, by the simple expedient of rectifying the installation defects and faults of the existing structure. Poor workmanship, inadequate detailing and lack of supervision can give rise to air gaps, broken seals and missing joints which may be the cause of the problem. Unauthorized variations by the original builder or a subsequent occupier may also be the source of problems that can be easily rectified. In older buildings wear and tear and minor movements may cause small gaps to appear which degrade sound insulation.

All such defects, particularly those that are readily identifiable and accessible, are remedied before a decision is made on a scheme of improvement works. It may be that this work will achieve a sufficient improvement to the insulation, and in any event it will ensure that the insulation provided by the structure is measured and not the transmittance of a number of small air gaps or ill-fitting doors.

Once a proper assessment of the problem has been made, and the need to improve the insulation of the building structure demonstrated, consideration can be given to suitable forms of treatment.

In the first instance select one or more of the techniques available which, even with the limitations of remedial works, should give substantial improvements and achieve an acceptable or reasonable level of sound insulation. The precise choice of treatment will be dictated by the particular constraints of the individual project.

From experience of remedial schemes, and the usual constraints, it is possible to highlight the more practical and reliable treatments.

Floors

Whatever treatment is adopted it should be applied to all party floors and extend into fitted cupboards, built-in wardrobes and other spaces.

Timber floors   Where there is sufficient ceiling height the secondary independent ceiling (Figure 4.6) is a reliable and effective technique for improving the sound insulation of traditional timber floors.

The original ceiling must be put into reasonable repair. (that is, no holes or gaps) and the electric wiring for the room lighting must be extended to the new ceiling. If the floors of the room above are formed from plane edged board they should be covered with a layer of hardboard to seal all air gaps (this is a worthwhile precaution whatever the original boarding). This technique does not disturb floor levels and can be carried out in occupied dwellings with a minimum of disturbance.

Where ceiling heights do not allow the use of the secondary independent ceiling the use of a platform floor (Figure 4.4) should be considered. This treatment causes more inconvenience to occupants, especially where works to both sides of the floor-ceiling structure are required (it is likely that some occupants on the 'floor' side may require temporary rehousing). A platform floor also changes the floor level by up to 50mm; all doors will need to be trimmed and rehung and in some cases it may be necessary to reposition kitchen or bathroom fittings.

In situations where neither the secondary independent ceiling nor the platform floor is practical the engineer should use the rafted floor with resiliently mounted ceiling (Figure 4.5). This causes about the same inconvenience to occupiers and is more complex to construct, but does not significantly change floor or ceiling levels. The detail (Figure 4.5) may be used where the existing ceiling is of insufficient mass but in otherwise good condition.

Other forms of treatment are inadvisable as they may add too much mass for the existing structure (for example heavy pugging) or there are insufficient data on their reliability.

Concrete floors   Usually existing concrete floors are a problem only in respect of impact sound insulation. Remedial works will be determined by the mass of the existing floor and, where the floor is heavy concrete, treatment may be by provision of a suitable resilient finish.

For medium and light concrete floors a floating floor (Figure 4.3) should be specified. For light or very light concrete floors that will not support the additional mass of a floating screed, it is possible to use a timber floating floor combined with a secondary independent or resiliently mounted ceiling. Apart from the simple provision of a resilient floor finish it is unlikely that remedial works to concrete floors can be carried out with the (upper) dwelling occupied.

Stairs Where stairs need to be provided with improved sound insulation it is usually in respect of impact noise. Stair treads should be treated with a resilient finish: for concrete stairs this can be the same material as used for floors; for timber stairs this should be a similar specification to a platform floor except that only a single layer of board need be used (adjoining floors, landings and so on should be treated in the same manner to equalize levels).

Airborne sound insulation between stairwells and adjoining dwellings should be controlled by the separating walls. In the rare example of a timber stairway soffit forming part of a dwelling, the soffitt should be treated for airborne noise using a modified form of timber floor treatment.

Walls

Assuming that the wall has been constructed as designed, the problem is reduced to determining whether it is practical to increase the insulation of the wall by the addition of mass and/or the provision of an additional independent leaf.

Masonry walls To increase substantially the insulation of an existing masonry wall by the direct addition of mass, considerable extra mass will require to be added. Under ideal conditions the mass law predicts that a doubling of mass is required for an average increase in the insulation of the order of 6 dB. The mass of the wall will need to be increased by a factor of four to achieve a 12 dB improvement in insulation.

For a 112mm-thick (half-brick) wall, regularly encountered as brick and block internal walls, it is unlikely that an acceptable method could be found to support the additional mass required. It is also unlikely that such an improvement would be accepted by the occupants, since the reconstruction of the party walls would create excessive disturbance.

For such walls it is usual to specify a remedial scheme of works that utilizes secondary independent panels (Figure 4.7) on both sides of the wall to improve the insulation. Where only one side of the wall can be treated (the other side may be a corridor already at minimum width), the mass of the panel should be at least 40kg/m² and the gap between the panel and the wall should be at least 100mm and ideally 150mm.

If this encroachment into the room is unacceptable the panel or its studs may be fixed to the existing wall using resilient bars, giving an added thickness of only 65–85mm, but at the expense of some of the possible improvement in insulation. This particular treatment will thus be unlikely to achieve a reasonable standard of insulation. For this reason its widespread use cannot be recommended, but in certain situations it may be the only practical treatment available.

Any electric outlet sockets or switches mounted on the original wall should be repositioned on another wall. Once the electrical fittings have been removed all spaces, voids and conduit left behind must be filled before the new panel is installed.

Timber-frame walls Timber-frame walls, whether structural frames or simple studs, are of lightweight construction. It is possible to achieve worthwhile improvements by the addition of a reasonable additional mass as doubling or even trebling the thickness of the wall cladding is usually within loading constraints. Such an increase to an existing structural frame, with good isolation from its neighbour, will frequently achieve a reasonable level of insulation. Where a structural frame does not have good isolation, or where the wall is of simple stud construction, it is unlikely that addition of mass alone will provide a solution. It will be necessary to introduce isolation and structural discontinuities usually in the form of a secondary independent stud wall. As with brick walls the depth required for the treatment can be reduced by mounting direct on to the existing wall with resilient bars but again with a reduction in the insulation improvement. The electrical fittings must be repositioned and all resulting holes must be filled.

Most stud walls are built directly on the floor joists, and the joists themselves are continuous through the floor structure on both sides of the wall. In this situation, even if the insulation of the floor is not suspect, treatment should be required to the floor surface. The treatment may be either a resilient layer or a full floating floor, which will assist in increasing isolation between the two rooms and improve the isolation of the stud structure.

Compound walls An example of a compound wall is where a doorway or disused opening has been blocked in at some time during an earlier refurbishment or alteration. The construction used to infill the opening may not have as good insulation as the surrounding wall, for example stud work may have been used to infill a disused doorway in a masonry wall. The presence of compound walls may not at first be readily apparent so check all likely locations of former openings. If such an arrangement is found the infill section should be stripped out and replaced with the same form of construction as the surrounding wall.

Services

In a scheme of remedial works some service installations may be disturbed by the works required to upgrade the insulation of party structures (for example back-to-back installations of electrical fittings). The opportunity should be taken to reposition or relocate these services so that they cannot affect the sound insulation. Other services and service runs will usually be left undisturbed or simply be enclosed in secondary ducting if they are likely to provide a transmission path for noise.

Trouble shooting

Trouble shooting is ensuring that appropriate sound insulation levels are realized by identifying and preventing possible failures. In respect of new-build or rehabilitation schemes it will include ensuring that the project is constructed in accordance with the agreed design and the insulation targets are achieved. Where the project is to remedy inadequate insulation in an existing dwelling the task will be to identify the cause of the problem, specify an appropriate scheme of remedial works and ensure that the remedial scheme is installed properly and the design targets met.

Common failures

Design failures are frequently caused by inadequate acoustic assessment at the design stage. This may result from the design team not having sufficient acoustic knowledge or specifying inadequate control measures and forms of construction. A design failure may also result from *ad hoc* changes in the design, particularly at a late stage of the project, which alter the acoustic requirements.

Construction failures occur where individual elements fail to provide suitable insulation because the expected level of insulation is not achieved, or is achieved but is still inadequate. The former may be the result of poor workmanship, flanking transmissions or unauthorized variations in materials and detail. The latter may be due to a design failure or because it is not practical to attain a higher level of insulation.

Preventing and remedying failures

Pre-construction works

In situations where the design development took place without suitable acoustic advice or such advice was sought at a late stage of development a paper assessment of the design should be carried out. All aspects of the design, layout and detail which could give rise to problems should be identified and possible alternatives suggested. In extreme cases it may be necessary to advise the design team or client that a fundamental re-think is required before the scheme progresses any further.

During construction works

As work proceeds periodic inspections must be made to verify that the project is being constructed in accordance with the agreed design. Inspections should cover the layout of the building, the construction of individual elements, and the standard of workmanship. In any project it is inevitable that variations to the original design will be made during construction and each proposed variation should be assessed for its likely effect on sound insulation. Where necessary compensation to the designed sound insulation may have to be made there should be clearly defined responsibilities for supervision on a day-to-day basis and procedures for authorizing variations to the design.

Pre-occupation

On practical completion of the project, but before handover, a final inspection should be made to ensure that all necessary insulation works have been completed to a satisfactory standard.

A completion inspection should include measurement of the sound insulation and certification of the insulation to the architect or client. Where a completion inspection or verification measurement has revealed inadequate insulation, handover of the building should not be accepted. The cause of the inadequacy should be ascertained and rectified.

Post-occupation

The existence of inadequate sound insulation post-occupation is likely to be identified by the occupants themselves. All complaints should be followed up by inspection and assessment. A distinction must be made between complaints of noise resulting from insulation failure, those resulting from unreasonable behaviour, and complaints resulting from unreasonably high expectations.

Expected performance and cost

With all features of a building there is a trade-off between the level of performance and the cost of providing that performance. The 'trade-off formula' is markedly different for new-build and for remedial works. Some techniques which are economic for a new-build project will be impractical or uneconomic for remedial schemes. In some situations it may be impossible to achieve a desired level of insulation by conventional remedial works. A balance must be maintained between cost and performance to achieve the most cost-effective solution while still ensuring that the scheme meets minimum insulation requirements.

New-build and rehabilitation

The minimum requirement for new-build is a reasonable level of insulation (see Table 4.2), and in practice a good level should be seen as a desirable insulation target.

The cost of achieving a reasonable level of sound insulation should not be seen as an additional cost to the project. It is a basic requirement of the design in the same way as adequate heating or lighting. In practice a reasonable level of insulation can be achieved by careful selection of conventional construction techniques, layout design and adequate supervision during construction. The cost of the insulation is included in the cost of the building elements and their construction.

If a good level of sound insulation is the target, some small additional cost can be attributed to its provision. The additional cost may be in respect of labour and materials specifically for insulation improvement, or it may result from changes in layout design. For instance, the position of the main service runs may be changed solely because of insulation considerations or they may be ducted to prevent a transmission path.

With careful design work the additional cost of providing a good level of insulation, as opposed to a reasonable level, will be of the order of 2–4 per cent of the project cost. This order of cost increase is seen by most people to be justified to attain the higher level of protection afforded by good insulation.

Remedial works

The performance-cost relationship for remedial works is markedly different from that for new works (apart, of course, from the general principle that increased performance means increased cost). Due to the constraints of remedial works some types of treatment may not be practical. Some treatments may not be advisable because they will not rectify the problem or provide the desired level of insulation, and some defects may be impossible to remedy without substantial reconstruction. Different projects will require different improvements in insulation because the existing levels vary. The law of diminishing returns may mean the improvement of a case of intolerable insulation to reasonable appears more cost effective than an improvement from reasonable to good.

Certainly a minimum requirement of reasonable insulation should apply in all cases, and is achievable. Good insulation is a desirable target for remedial works but, unlike new-build, it may be impractical to achieve. However, where there are minor flanking transmissions and the basic layout design of the dwellings is satisfactory, good insulation is usually attainable.

Each case of remedial works needs to be costed as a unique exercise and the choice of treatment must be specific to the individual situation. For remedial works it is possible to give only examples of costs which have been encountered, and their relevance to other projects is a matter of judgement.

Consider a case of flats having inadequate insulation across the party floors.

Where there are no main flanking paths and where the floor is of conventional construction, the cost-performance shown in Table 4.3 may be expected.

Legal considerations

The legal considerations of improving sound insulation in buildings invariably fall into those relating to the prevention of potential problems or those relating to the remedy of known problems. There is also the legal problem of financial responsibility, which is frequently the source of a quite separate dispute. For the sake of simplicity only the more regularly encountered legal provisions will be considered here, particular emphasis being given to those problems or issues which affect residential properties.

Statutory controls relate, in the main, to residential property; also the bulk of case law and legal experience in this field has arisen as a result of actions relating to residential usage.

The legislative control outlined in the following sections apply to the British Isles, but the discussion should be of benefit, or interest, to all readers.

Prevention

In considering the prevention of problems caused by inadequate sound insulation, a distinction must be drawn between buildings which are constructed completely anew (new-build) and those existing structures which have been given a new or changed use as a result of improvement and reconstruction (rehabilitation).

This distinction is necessary because legislative controls frequently

Table 4.3
Typical cost-performance data for remedial works

Original construction		Treatment	Performance Before	After	Cost* (£/m²)
Suspended timber	1	Secondary independent ceiling	Poor	Good	45–55
	2	Rafted floor with resilient ceiling	Poor	Good	38–50
Concrete slab	1	Resilient finish	Poor†	Good	25–40
	2	Floating screed	Poor	Good	35–45

*Approximate costings at 1987/8 prices, upper limit relates to work in occupied dwellings.
†Impact only.

distinguish between new-build and rehabilitation, and the two situations may be treated differently.

New-build

Most countries now have well-established legal requirements governing the construction of new buildings. In England and Wales there are the Building Regulations 1985,[12] Scotland and Northern Ireland also having similar legal provisions.[16,17] In addition to the various national codes it is also possible to find state or city ordinances, particularly in countries where there is a well-defined federal structure.

In England and Wales the Building Regulations seek to control the manner in which construction work is carried out, the quality of materials used and the design of structures. The provisions of the regulations are enforced by local authorities, and since 1985 apply nationally. The general legal provisions are contained in the regulations themselves, while the technical requirements necessary to comply with the regulations are contained within supporting codes known as Approved Documents. The technical requirements relating to sound insulation are contained in Approved Document E.[13]

Proposals for new buildings must show compliance with the regulations at the design stage in order to gain building permission. The design is scrutinized, usually by the local authority, which if satisfied will issue an approval under the regulations. During the building works periodic inspections should be made by the local authority to verify that the construction has been carried out in accordance with the approved design.

Any measurements or performance figures should be conducted and quoted in accordance with the associated British Standards.[3,5,8]

Rehabilitation

In some countries, Scotland is an example, the legal building controls will also apply to rehabilitation schemes. New party structures or existing partitions that become party structures will be expected to meet the insulation requirements of the relevant building codes. Plans must show the means of upgrading the insulation of existing elements before formal approval will be given.

Where rehabilitation schemes are exempted from the insulation requirements of building control legislation, as is the case in England and Wales, new dwellings can be created which do not have acceptable sound insulation. Local authorities wishing to prevent this situation occurring will usually rely on their powers under planning legislation.[14,18]

A rehabilitation scheme which converts a large building into several new dwellings will require planning permission from the local authority. The local authority may make its approval for the scheme conditional upon acceptable sound insulation being included. A condition may take the form of a performance standard, or it may require a specific construction technique. Some authorities impose a requirement that the developer must submit details of the scheme of insulation to be adopted. In practice, the conditions seek to achieve an equivalent insulation performance to that expected by the Building Regulations.

Unlike the national code of the Building Regulations the use of planning conditions in this way is at the discretion of the individual local authority. As result there are both policy and procedural variations between different local authorities. It should be noted that at the time of writing this situation is under review nationally.

Remedy

Nuisance

Assuming that a noise problem can be attributed to inadequate sound insulation the basis for liability and action is usually that of nuisance.[18-20] The concept of nuisance, and the factors influencing it, are contained within the common law as established by case law and precedent. The legal remedy of a noise nuisance is usually pursued via either the common law of nuisance or the Control of Pollution Act 1974.[19]

Common law   The essence of legal nuisance is that it is a condition or activity which unreasonably interferes with a person's rights or use of land. Nuisance is best understood by such terms as unreasonableness, annoyance, inconvenience, interference and the like. Where the nuisance is the result of inadequate sound insulation of the building structure, the person responsible for that structure (usually the landlord or freeholder) may also be the person responsible for the nuisance. Under common law the persons suffering the nuisance may take an action against the person responsible and secure a remedy in the courts. The remedy will usually be damages, not remedial works, based on the inconvenience suffered.

Control of Pollution Act 1974   This Act makes noise nuisances statutory nuisances, that is, a noise nuisance may be dealt with summarily using the provisions laid down in the Act. The test of nuisance is the same as that used under common law and action is taken against the person responsible for the nuisance.

Unlike common law, action using the provisions of the Control of Pollution Act can be taken by either an individual suffering the nuisance or the local authority (but there are slight procedural differences). In fact, once a local authority is satisfied that a nuisance exists the Act places a mandatory requirement on the authority to take action. The remedy under the Act can be a fine imposed by the magistrates' court or, more likely for sound insulation problems, a scheme of remedial works which will abate the nuisance.

Successful nuisance actions in respect of inadequate sound insulation have been taken under all the provisions described above, and it is interesting to note that the procedures are not mutually exclusive. It would be possible, say, having achieved abatement using the provisions of the Control of Pollution Act to take a civil action for damages (though the damages must be sought by the individual who had suffered the nuisance) covering the period up to the abatement of the nuisance.

The proof of the existence of a nuisance will not necessarily require complex measurements of the insulation or noise levels. Nuisance is a subjective condition which does not lend itself to definition by objective measurements. A condition or activity which can be considered a nuisance in one situation may not be nuisance in another.

The detailed legal and procedural points relevant to nuisance are outside the scope of this book. Readers wishing to study the subject in more detail should refer to works which deal generally with the law relating to noise,[21] or specifically with the legal problems of inadequate sound insulation.[18,22]

References

1   Scholes, W.E. and Jones, A.J. (1984) Sound Insulation between Dwellings: The Validity of National Performance Standards, *London Environmental Supplement*, No. 9, Winter.

2   British Standard Code of Practice CP3: Chapter III (1972) *Sound Insulation and Noise Reduction*, British Standards Insitute.

3   BS 2750 (1980) *Methods of Measurement of Sound Insulation in Buildings and of Building Elements*, Part 4: Field measurements of airborne sound insulation between rooms; Part 7: Field measurements of impact sound insulation of floors, British Standards Institute.

4   ISO 140/IV/VII (1978) *Acoustics – Measurement of Sound Insulation in Buildings and Building Elements*, ISO.

5   BS 5821 (1984) *British Standard Method for Rating the Sound Insulation in Buildings and Building Elements*, Part 1: Method for rating the airborne sound insulation in buildings and of internal building elements, British Standards Institute.

6   ISO 717/I (1982) *Rating of Sound Insulation of Dwellings*, Part 1: Airborne sound insulation in buildings and of internal building elements. ISO.

7   ASTM E 413–73 (1973) *Standard Classification for Determination of Sound Transmission Class*.

8   BS 5821 (1984) *British Standard Method for Rating the Sound Insulation in Buildings and Building Elements*, Part 2: Method for rating the impact sound insulation, British Standards Institute.

9   ISO 717/II (1982) *Rating of Sound Insulation of Dwellings*, Part 2: Impact sound insulation, ISO.

10   ASTM E 492–77 (1977) *Laboratory Measurement of Impact Sound Transmission through Floor – Ceiling Assemblies Using Tapping Machines*.

11 FHA No. 750 (1963) *Impact Noise Control in Multi-Family Dwellings*, Federal Housing Administration.
12 Building Regulations 1985, Part E, Schedule 1, HMSO, London.
13 Approved Document E/1/2/3 (1985) *Airborne and Impact Sound*, HMSO London.
14 National Society for Clean Air, Noise Committee (1986) *Report on Sound Insulation in Flat Conversions*, Part 1: Outline of the problem and NSCA survey results.
15 The Housing Corporation (1985) *Design and Contract Criteria*, Issue 3/3, London.
16 Building Standards (Scotland) (Consolidation) Regulations 1971, Part H: Resistance to the transmission of sound, HMSO, London.
17 Building Regulations (Northern Ireland) 1977, Part G: HMSO, London.
18 National Society for Clean Air, Noise Committee (1987) *Report on Sound Insulation in Flat Conversions*, Part 2: Legislative and technical solutions.
19 Control of Pollution Act 1974, HMSO, London.
20 Bassett, W.H. (1983) *Environmental Health Procedures*, Lewis, London, pp. 113–17.
21 Kerse, C.S. (1975) *The Law Relating to Noise*, Oyez, London.
22 Rintoul, S. (1986) Some Economic and Legal Considerations of Improving the Sound Insulation of Party Floors in Converted Dwellings, *Proc. IOA*, Vol. 8, Part 1, pp. 9–16.

Chapter 5

NOISE CONTROL WITHIN THE INDUSTRIAL ENVIRONMENT

John Roberts and **Bridget Shield**, Acoustics Group, Institute of Environmental Engineering, South Bank Polytechnic.

This chapter discusses general methods of reducing noise levels inside the workplace. The problems of individual hearing protection are discussed in detail in Chapter 6.

Reasons for noise control

Exposure to excessive noise levels at work damages the hearing of employees. A short exposure can cause temporary hearing loss for a period lasting from a few seconds to several days. Regular exposure to such levels over a long period of time can result in serious loss of hearing which is permanent and incurable.

In addition, noise may interfere with working efficiency and disturb concentration, especially where work is difficult or highly skilled. It may also hinder communication, mask warning signals and so contribute to accidents.

The 1986 European Council Directive 86/188/EEC[1] applies to noise hazards and provides for the possibility of specifying maximum noise levels and introducing other measures to protect the estimated five million workers at risk from exposure to noise. While this Directive does not prejudice the right of member states to apply their own laws and regulations, the adoption of a recommended sound exposure level by the EEC is an important impetus for the reduction of noise levels in industry. In the UK, where the *Code of Practice for Reducing the Exposure of Employed Persons to Noise*[2] recommends that workplace levels do not exceed 90 dB(A) L_{Aeq} (8 hour), claims against employers for noise-induced hearing loss are currently being settled at around £10 000, depending upon the severity of the injury sustained. One industrial insurance group has set aside a special fund of £5 million to meet expected future claims.

An integrated approach

An integrated approach to noise control is essential for cost-effective long-term solutions, both at the design stage and where remedial measures are taken. Extensive experience in Europe and the USA has shown that an integrated programme of noise control costs substantially less than piecemeal *ad hoc* exercises. For example, with proper pre-planning British Gas satisfactorily quietened a compressor station for 2 per cent of the total costs whereas the cost of remedial treatment to an equivalent French station rose to 10 per cent of the total cost.

Noise control at the design stage

In the early stages of any project in which noise control may become a significant element a qualified acoustician should be an integral member of the design team. Such an arrangement ensures consideration of noise control at all stages of development and enables effective communication with associated engineering disciplines.

The cornerstone of any noise control programme is the determination and adoption of design objectives. For workplaces noise levels of less than 90 dB(A) L_{Aeq} (8 hour) should be aimed at, and for other areas target levels may be found in appropriate design guides, for example those of the Chartered Institution of Building Services Engineers[3] or the American Society of Heating, Refrigeration and Airconditioning Engineers.[4]

The most cost-effective noise control recommendations are likely to be made during the preliminary stages of design. Working directly with mechanical, electrical or building services engineers, the noise control consultant can often define and alleviate potential noise problems early in the project. Factors which should be considered include plant layout, the selection of building materials and the choice of equipment and services.

Building layout

Noise problems may often be reduced, and sometimes eliminated, by modifications to the proposed building layout. The relative placement of noisy and quiet areas in any given scheme should take account of the following guidelines:

1   Quiet rooms should be located as far as possible (vertically as well as laterally) from noisy rooms or external sources of noise.

2 The windows in quiet rooms should be arranged to face away from potential noise sources.
3 Quiet rooms should be defended from noise by corridors or rooms with less stringent noise criteria.
4 Noisy rooms should be grouped together so that their effect is not so widespread.
5 Powerful noise sources should be enclosed by constructions of sufficient mass to reduce external levels to acceptable values.

Figure 5.1 shows an example of obviously good and bad layouts for a typical teaching department.

Building materials

Another element of noise control important in the design stage is the correct selection of building materials. In one factory a predicted reduction in sound level of 4 dB was obtained for a notional increase in cost by changing the room finish. Low-cost treatment might include not painting such finishes as fibreboard tiles, to avoid reducing the acoustic absorption coefficient.

Equipment and services

As the project progresses acoustic performance must be included in specifications for all mechanical and electrical equipment and services. The quietest version available within the budget should be chosen, with any increase in cost balanced against the technical feasibility and cost of on-site engineering solutions required to deal with residual noise problems.

The most direct and convenient method of ensuring that a machine is not excessively noisy is to specify maximum permitted noise levels in the purchase agreement, but it is important to appreciate that conditions of use may dramatically alter noise output. An item of equipment properly mounted and running under the near ideal conditions of a manufacturer's test bed may not be so quiet when in the factory.

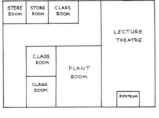

BAD LAYOUT

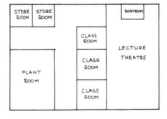

GOOD LAYOUT

Figure 5.1 *Examples of good and bad layouts for a typical teaching department*

For example, one large extract fan delivered as giving 85 dB(A) at one metre produced a sound level of 108 dB(A) after installation. It had been mounted on a metal gantry with grossly inadequate vibration isolation, and with inlet and outlet conditions that could hardly have been better designed to produce turbulent, noisy flow. This is a good illustration of inadequate collaboration and failure to appreciate the importance of the acoustic dimension until too late.

With the installation of plant and machinery structureborne noise is a frequent occurrence unless the equipment is adequately isolated from the structure. Flexible couplings and anti-vibration mounts should be specified for all items fixed to floor slabs or hung from ceilings. Pipework and ducting through walls should be suitably sleeved with no air gaps, and radiators of noise such as fan extracts placed where the noise emitted will not reach noise-sensitive areas.

It is essential that a series of on-site noise level verification measurements is made both at the commissioning stage and during commercial operation.

Remedial noise control

Notwithstanding the introduction of noise control measures from the earliest stages of a project a common application of noise reduction techniques is remedial, concerning *in situ* palliative actions.

A satisfactory solution to any industrial problem must be cost effective and founded on a knowledge of the manufacturing process or service. For example, fitting a commercially available enclosure around a woodworking machine may overcome the noise problem quite simply, but fitting such an enclosure without consideration of the method by which the wood is fed into or exits from the machine, or the manner in which the operator controls the machine, may result in excessive delays or a lowering of the quality of the work output.

Such is the wide diversity of plant, machinery and processes used in industry today that it is impossible in a few pages to review every aspect of noise control. Those noise reduction techniques which have been found most effective are outlined followed by illustrative examples which include estimates of the costs involved.

The three basic elements of any noise problem are the noise source, the path taken by the sound and the receiver. The general approaches to the reduction of unwanted sound are, in order of likely efficiency and cost-effectiveness:

1   Reduction of noise at source, including the redesign or replacement of machines, or alteration of the manufacturing process.
2   Placing an acoustic barrier between source and receiver, or erecting a total or partial enclosure around either.

3 Using acoustic absorbent within the space to reduce the general background or reverberant noise field.

In addition, alteration of work practices and proper maintenance of machinery may be used to reduce the noise dose received by any particular group of workers.

Where there is more than one noise source present it is normally essential to quieten the noisiest source first, followed by other sources, if there is to be any substantial reduction in the overall sound level.

General acoustic principles may be used as a guide for noise control, but the complexities of industrial noise sources and their environments make it difficult to predict quantitative results after treatment. Each noise control application needs to be analysed within its own mechanical, operational and environmental context.

Assessing the problem

The first step is to define fully the extent and magnitude of the problem. This will necessitate measuring the relevant noise spectrums and dB(A) levels and securing as much information as possible on the environment in which the problem exists. This includes the appropriate criteria for the particular situation and complete data on all noise sources and equipment. The reduction in noise level necessary to meet the agreed criterion can then be determined.

After the required degree of noise reduction has been established the various methods available for reducing the particular type of noise should be evaluated. The two key elements in such an appraisal are an estimate of the likely results and the cost of implementation of the noise reduction programme including any disruption of work schedules.

Reduction of noise at source

The method of noise control to be used will depend on the type of sound that is causing the problem.

Impact sound

Impact noise from such sources as castings thrown or falling into skips is widely encountered and is commonly dealt with by using vibration dampening materials. Such materials must be extremely resistant to wear. Special forms of rubber (stuck or bolted on) have been successfully used as have sprayed coatings of urethane. These coatings should be at least as thick as the vibrating metal. Wooden linings have also been found to be effective but rapid wear and resultant cost are major disadvantages. Wooden linings also tend to splinter and may become oil soaked thus creating a fire hazard.

Any sound-deadening material placed on the impact surface eventually deteriorates and it is advisable to use it in a sandwich protected by a metal impact plate. Various materials have been used in this way to reduce the amplitude of vibration and hence the acoustical energy radiated. Typical materials include rubbers, sand, concrete, fibreglass and plastic foam. Inclusion of a layer of high-damping visco-elastic material can be very effective for thinnish sheets of relatively large surface area, but is very much less effective for thicker sections.

The use of a layer of damping material can be extended to any thin sheet-metal surface. In many cases the material used is mineral wool blanket 30 to 50mm thick, spot glued to the surfaces and covered with an outer protection of suitably perforated sheet metal. Such an arrangement is useful for, say, electrical control cabinets or plate machine surfaces and has the additional advantage of increasing the acoustic scattering and absorption properties of the item, as described later in this chapter.

These methods all give noise reductions, but serious problems of this nature are best overcome by using non-metallic containers. In one case a metal exit chute produced peak noise levels of over 100 dB when components were dropped into it.[5] Replacing the metal chute with one of ultra-high molecular weight polythene reduced the peak levels to about 86 dB, and the L_{Aeq} level by about 10 dB(A). In addition the life of the chute was increased, the replacement cost was halved, wear did not produce sharp, dangerous edges on the chute and component damage was reduced.

For the particular situation where grinding and cutting thin metal sheets can be very noisy, such as the finishing of a water tank, elastic magnetic plastic mats are available and may reduce noise levels by as much as 10 dB, though mainly at high frequencies.

Where impact forces are unavoidable it is important to reduce their amplitude to a minimum by, for example, the use of resilient buffers. The amplitude of vibration should also be limited by the use of an adequately rigid or clamped frame. The vibrating area may be minimized by the inclusion of breaks to prevent the passage of vibration.

Impact processes are always noisy so alternative processes should be considered at the initial stage. For example, riveting by compression and compacting moulds or deforming sheet metal by hydraulic pressure are obvious alternatives. There are many more.

Aerodynamic noise

A common and often serious source of production noise is aerodynamic noise generated by pneumatic discharge systems, blow-off nozzles, steam vents and leaking high-pressure air lines.

Noise levels from compressed air exhausts are generally easily controlled by commercially available mufflers giving reductions of up to 20 dB(A). Figure 5.2 shows a design which has proved especially good at reducing general

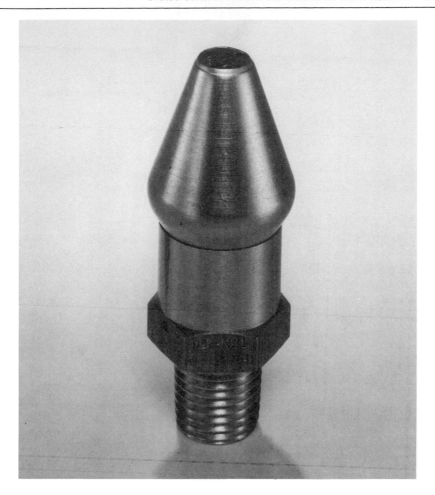

Figure 5.2   *High pressure exhaust muffler*
(Reproduced by permission of Barry Controls Limited.)

noise levels from high pressure exhausts. The noise generated by an air jet strongly depends upon the velocity gradient between the jet and the surrounding medium. Introducing a secondary airflow around the jet and so reducing the sharpness of the velocity gradient (as with a by-pass jet engine) is a common and successful method.

Simple durable silencers can be constructed by maintenance personnel using ordinary pipe fittings and filler of brass wool or some other non-rusting material, as shown in Figure 5.3. Up to 10 dB(A) reduction can be obtained in this way. The problems associated with such silencers are that they generate a backpressure and may be of such a size as to interfere with normal operation, but their very low cost and ease of fitting are decided advantages.

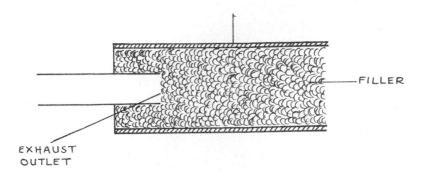

Figure 5.3 *Silencer made from pipe fitting and filler*

Where the compressed air performs work after leaving the pipe network, such as cleaning sand from moulds, the use of multiport nozzles will alter the spectral shape and reduce the dB(A) levels by shifting the noise peak to a higher frequency. The reduction achieved depends upon the number of ports in the nozzle (generally the more the better) and the air pressure, but in certain high-velocity cases reductions of up to 20 dB(A) have been achieved by this method.

It is often best to conduct air discharges to a remote point outside the work area using a length of suitable heavy duty pipe, taking due care to avoid causing annoyance to the public. Substantial noise reductions have been achieved in the workplace using this method, but it is advisable to filter the air to avoid blocking the output silencer with oil or swarf.

Supply and extract fans are a main cause of noise nuisance. All fans are noisy, generating noise in a variety of ways, both broad-band noise from air turbulence within the fan casing and discrete tones due to the interaction of impeller blades with vanes or casing. This noise radiates directly into the airstream where it exits more or less equally from fan inlet and outlet. The control of fan noise is discussed in Chapter 7.

Noise from industrial combustion systems

Noise from boiler plant can be divided into four general categories, each requiring its own methods of noise reduction. In order of likelihood of occurrence these are:

1 Fan and/or booster noise. The fan supplies the air for forced draught combustion and the booster raises gas pressure to the burner. These noise problems are usually solved by placing commercially available shrouds over the offending items, as shown in Figure 5.4. At least 10 dB(A) noise reduction is achievable by this method.

Figure 5.4 *Burner silencer*
(Reproduced by permission of Weishaupt (UK) Limited).

2 Wide-band noise originating from the 'roar' of the flame. There may be an acoustic weak path allowing noise to escape from the combustion chamber. If such a path exists it will probably be at or near the burner, or the noise may even be radiating from the burner itself. A shroud or fibreglass cladding faced with aluminium placed over the problem areas may reduce the noise to satisfactory levels. Burner operation should be checked to ensure that excess air is the minimum acceptable, say 10 per cent. Reducing throughput will invariably reduce noise levels and this will often be possible since most combustion systems tend to be oversized. If these measures are insufficient expert advice should be taken.

3 Pump noise. This may be alleviated by providing adequate vibration breaks, including flexible couplings, and, if necessary, small acoustic enclosures around the pumps.

4 Combustion-driven resonance. Such a noise problem implies acoustic coupling between the flame and a resonant mode of the combustion chamber/flue system. Helmholtz resonators, high-temperature acoustic absorbent, improved flame retention, active control devices and even a hole knocked in a chimney have been used as cures. Expert advice should be sought.

The plant room containing the heating plant needs to be of sufficiently massive construction to reduce transmission to the exterior to acceptable levels. Combustion processes need air, the openings for which generally require acoustic louvres and should be positioned to cause minimum external noise problems.

Reduction of noise by enclosures, partitions and screens

By far the most common method of reducing industrial noise, and still generally the most cost effective in practice, is the use of acoustic barriers. These include complete or partial enclosures around either source or receiver, partitions or screens in the acoustic path, and ear defenders worn by the workforce.

Enclosures

When control of machine noise at source is impracticable an enclosure may be used to reduce the sound. It is sometimes necessary to surround the entire machine by a free-standing enclosure, while in other cases it is possible to reduce the noise sufficiently by enclosing only the noisy components of the machine. Enclosures are a popular means of noise reduction because of their relative ease of construction and fitting, but their acoustic behaviour is complex and difficult to analyse. The prediction of the performance of enclosures therefore tends to be governed by empirical rules, common sense and experience.

Complete enclosures

Commercially available enclosures as shown in Figure 5.5 exist to meet virtually all needs. These generally perform well, look good and are not necessarily expensive. A principal advantage is that the sound reduction achievable with a commercially supplied enclosure is known to a fair degree of accuracy.

It is well worth perusing the available literature before building on-site acoustic enclosures, and it should be noted that certain situations will require specialist advice, for example where backpressure across inlet and exhaust silencers may have an appreciable effect on the performance of the enclosed plant.

Enclosures built on site using available labour, simple tools and materials to hand (usually wood) can cure almost any noise problem. They offer the advantages of cheapness and direct control of the operation but care must be taken to ensure that there is no infringement of fire regulations. There may also be problems of size, appearance and durability.

Construction   The shape and density of the materials used to construct enclosures determine the amount of sound transmitted to the outside. Dense materials such as metal, wood, brick or concrete reduce transmitted noise and should be used to build the outer shell of the enclosure. The inside should be lined with acoustic absorbent consisting of fibrous materials such as acoustic board

Figure 5.5 *Acoustic enclosure*
(Reproduced by permission of Acoustic Treatments Limited.)

or mineral wool. This reduces noise levels within the enclosure thereby improving both the aural comfort of any operators working inside and the general reduction of transmitted sound.

When enclosing any machine consideration must be given to the provision of cooling and/or ventilation to prevent overheating and resultant damage to moving parts. Care must be taken to prevent the egress of noise from any such openings. Suitable intake and discharge silencers are readily available but if necessary air may be ducted to a place remote from the workforce.

Adequate vibration breaks should be provided to ensure that structureborne sound is not transmitted from the machine to the enclosure. Radiation of sound from the enclosure may also be reduced by ensuring that it is adequately damped to prevent resonances of its panels. This is usually achieved by lining with sound-absorbing material as described above.

Access   When it is necessary to obtain access to the machine, for example for maintenance or machine setting, the enclosure should contain doors or removable panels. These should be suitably robust and close-fitting, and held in place by quick-release fastenings. Experience shows that convenience of access is important; should the acoustic enclosure increase maintenance or setting times, access panels will quickly be 'lost'. Where the safety of the workforce is involved the doors and panels should be appropriately linked with safety switches.

Prediction of performance   In most cases it is a relatively simple matter to design and construct a suitable enclosure on site. The performance of the enclosure (required reduction in sound levels) is found by comparing the measured level with the criterion being used. The essential characteristic of any acoustic enclosure is its *sound reduction index* or *sound transmission loss*, *R,* for each relevant octave band. Values of *R* are obtainable from standard sources such as the National Physical Laboratory[6] or manufacturers' data.[7]

The values of *R* obtained from the reference literature are those measured under the idealized conditions of a laboratory and hence tend to be the maximum achievable values. In practice it is not generally possible to achieve such high values without the inclusion of absorptive linings. For example, a simple, rectangular, unlined, 6mm-thick plywood enclosure had a predicted average sound reduction index of 17 dB, but on-site measurements gave a value nearer to 10 dB. When the enclosure was lined with 50mm of mineral wool the measured value of the sound reduction index rose to very nearly the reference value.

In calculating the performance of an enclosure it is usual to consider only the reverberant or diffuse sound field and to assume that sound is transmitted uniformly through the enclosure. Such conditions are met only with larger enclosures and where there is sufficient distance between the surface of the machine and the enclosure to avoid acoustic coupling. The reduction in sound pressure level in the receiving room due to the insertion of the enclosure is termed the *insertion loss* (IL) and may be predicted approximately from:

$$IL = R - 10\lg(1/\alpha) \quad dB \tag{5.1}$$

where R is the sound reduction index of the panels of the enclosure, and α is the average sound absorption coefficient of the inside surfaces of the enclosure at the relevant frequency.

The likely reverberant sound pressure level, L_p, in a space due to the installation of an enclosed source whose sound power is known, can be predicted using the following formula:

$$L_p = L_w + 10\lg(4/R_c) - IL \quad dB \text{ re } 20 \ \mu Pa \tag{5.2}$$

where L_w is the sound power level of the source in the octave band of interest
R_c is the room constant of the space into which the enclosure radiates
($R_c = Sa_1/(1 - a_1)$ where S is the total surface area of the space and a_1 is the average absorption coefficient).

These two equations do not allow for direct sound radiation and so care must be taken in their interpretation.

Illustrative example   The 1 kHz octave band reverberant sound pressure level in a factory space due to a compressor is 80 dB re 20 μPa. The machine is $1 \times 2 \times 1$m high and is to have an enclosure $2 \times 3 \times 1.5$m high placed around it. The machine is on a concrete floor (absorption coefficient 0.04) and the interior surfaces of the enclosure are to be faced with 50mm-thick mineral wool blanket having an absorption coefficient of 0.75 at 1 kHz.

What should be the sound reduction index of the enclosure to reduce the external reverberant field due to the machine to NR 45?

To achieve a level of NR 45 the sound level in the 1 kHz octave band must be reduced to 45 dB. The insertion loss required is therefore 80 − 45 = 35 dB.

The total internal surface area of the enclosure is

$$2 \times [(2 \times 3) + (1.5 \times 2) + (1.5 \times 3)] = 27m^2$$

The total internal acoustic absorption of the enclosure is

$$6 \times 0.04 + 21 \times 0.75 = 15.99m^2$$

The average absorption coefficient, α, for the interior surfaces of the enclosure is thus given by, $\alpha = 15.99/27 = 0.59$.

From equation 5.1

$$35 = R - 10\lg(1/0.59)$$

$$R = 37 \text{ dB}$$

Thus the required sound reduction index for the enclosure is at least 37 dB at 1 kHz.

Alternative formulae   The presence in enclosures of reinforcing ribs, unknown damping factors, multiplicity of end fixings and conditions, and the variation in shape mean that the calculated value of R should be taken only as a guide. It is impossible to predict the insertion loss to better than several decibels. The standard formulae for sound pressure level tend to implicitly assume the numerical equivalence of sound pressure and sound intensity levels. This is true for plane waves but not for reverberant sound fields. In this latter case the reading on a sound level meter will overestimate the intensity level falling on the boundary surfaces by up to 6 dB. It should therefore be borne in mind that when the acoustic conditions on the two sides of the enclosure or partition are not both 'dead' or 'live', formulae such as 5.2 above can be in error by up to 6 dB. The worst case is when the source room is acoustically 'dead' and the receiving room highly reverberant, which tends to be the condition for acoustic enclosures and so it is usual to err on the side of caution. Many acoustic consultants therefore use the following modification of equation 5.1.

$$\text{IL} = R - 10\lg(1/a) - 6 \quad \text{dB} \tag{5.3}$$

In the given example the calculated value of the required insertion loss would then be raised to 43 dB.

If the enclosure is sufficiently far from any walls or other reflecting surfaces the nearfield sound level, L_n close to the enclosure may be approximated by

$$L_n = L_w - R - 10\lg A_e \quad \text{dB re 20 } \mu\text{Pa} \tag{5.4}$$

where A_e is the total acoustic absorption within the enclosure.

The total sound field at any point will be the appropriate sum of the near and reverberant levels.

Mechanical isolation   A vital requirement for enclosures is complete mechanical isolation from the source machine. This may be achieved using commercially available vibration isolation mounts between the machine and the floor, and vibration breaks in the form of flexible connections for all pipes and ducts passing through the enclosure. Where practicable service conduits should be grouped together to pass through an aperture as small as possible and appropriate sealing techniques used, such as the cutting of sheets of lead-vinyl to fit snugly around the conduit both outside and inside the enclosure.

Effect of resonances   Two types of resonance effects can considerably reduce the efficacy of enclosures at certain frequencies.

The coincidence effect, when bending waves propagate in the walls of the enclosure, occurs at a critical frequency given by

$$f = 66(M/Eh^3)^{0.5} \quad \text{kHz} \tag{5.5}$$

where M is the superficial mass of the wall of the enclosure, kg/m²
E is the Young modulus of the material, N/m^2
h is the thickness, m.

Thus for a 100mm-thick concrete wall the critical frequency will occur in the 250 Hz octave band giving a dip in the sound reduction index in this band in the spectrum. For concrete, with relatively high internal damping, this dip is not severe but for other materials, such as steel, it might prove troublesome. The frequency of the coincidence dip may be estimated from data given by Bazley.[6]

Panel resonances also occur at certain frequencies, but no adequate relationship exists to predict them for the multiplicity of composite materials and fixings used in practice.

It should be emphasized that to avoid possible acoustic coupling between source and enclosure, which could excite such resonances and substantially reduce the insertion loss of the enclosure, the distance between the source and the enclosure should be at least half the wavelength of the lowest frequency of interest.

Case history 1   Tumblers in a small naval foundry gave a peak sound level of 109 dB(A) with a broad-band spectrum centered on the 3.15 kHz octave band. It was decided to reduce this level to, at most, 90 dB(A) using an enclosure. The size of the enclosure, 2x3x2m, was such as to leave 500mm clearance around the tumblers. Initial calculations indicated that a sound reduction index of at least 19 dB was required. After reference to tables of typical sound reduction indices for the materials to hand the design shown in Figure 5.6 was built.

The framework consisted of 50 × 100mm studs located at 450mm centres, the 100mm depth being filled with dense mineral wool. The inside surface facing the tumblers was a 25mm-thick layer of fibreglass blanket suitably

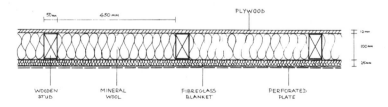

Figure 5.6   *Construction of wall of tumbler enclosure*

faced and covered with a perforated metal plate 2mm thick. The back of the enclosure was 12mm-thick plywood with angle iron at all corners. The access door was of the same construction with a 'pull-tight' handle. Apart from the handle all materials and labour were available on site so the add-on cost of the enclosure was marginal. After some months' usage the measured sound level with the enclosure in place was 87 dB(A).

It is instructive to appreciate that the effectiveness of the enclosure was less than that predicted from the bare design data. There were three reasons for this: the value of R used in the calculation was taken from reference tables and hence tended to be the maximum achievable; the effect of the floor and the electrical and mechanical connections were ignored; and it is usual in practice for wear and tear of doors and removable panels to cause some air gaps.

Figure 5.7   *Acoustic booth*
(Reproduced by permission of the Noise Control Centre.)

Alternative methods of enclosure

Acoustic booths   On occasions it will be necessary to reverse the procedures described above and enclose the workers in an acoustic booth as shown in Figure 5.7. Proprietary booths give excellent noise reductions of up to 25 dB(A) *in situ*. One such enclosure, measuring 10 × 4 × 2m cost £19 000 in 1987 for the shell and installation, with additional costs for air-conditioning, heating and lighting.

In designing such booths particular care must be paid to the general safety and comfort of the occupants and others. Considerations must include temperature, ventilation, lighting, visibility from the booth, occupancy and space requirements. The booth should be made inviting enough to encourage its use as a matter of choice.

The main difference between machine enclosures and acoustic booths is the need in the latter for windows. Ideally these should be double glazed. The gap between the two panes should be as wide as possible and at least 100mm. If different thicknesses of glass are used (say 6mm and 9mm) with airtight sealing into the structure and acoustically absorbent reveals, the windows need not be an acoustically weak point in the booth and there is no need to reduce their area to a minimum. If doors are used they should be solid core with adequate draught-proofing round their edges and 'pull-tight' handles. A typical booth of dimensions 3 × 3 × 2m, built on site in accordance with the guidelines above and using available labour and materials, gave 15 dB(A) reduction at a notional 1987 cost of under £2000.

Partial enclosures   Free-standing enclosures, which should be sealed around their base for best performance, tend to be permanent and may occupy valuable space. Where the noise output from a machine can be identified as radiating from certain specific components it is often advantageous to construct enclosures for those parts only.

The use of such partial enclosures can be just as effective as enclosing the whole machine provided all parts of the machine which radiate noise are covered. However, serious limitations are imposed on all such enclosures by the need not to interfere with operation, safety or maintenance.

Specific partial enclosures purpose-built for individual machines may be constructed by factory staff. It is advisable to test the efficacy of the proposed design using a mock-up. Corrugated cardboard lined with a 25mm layer of acoustic absorbent and held together with heavy tape will give a good indication of the attenuation likely to be achieved, and has the added advantage of enabling practical diagnosis of design faults at an early stage. One such mock enclosure around the die, ram, feed and chute to the storage bin of a punch press gave a sound reduction of 6 to 7 dB(A). The final plexiglass and metal enclosure gave 8 to 10 dB(A) reduction and as much as 18 dB in the 8 kHz octave band.

Partial enclosures tend to be widely used in the woodworking industry because the principal noise sources are obvious and readily defined. The difficulty is that the wood for the machines must be fed in and this means a gap in the enclosure through which the sound will escape. It is therefore important that these openings are as small as possible. They should take the form of lined ducts fitted with curtains of heavy rubber or PVC strips long enough to make an effective seal when not displaced. Such enclosures have given 5 to 15 dB(A) reductions depending upon the measurement position, the number of gaps and the peak frequency of the noise.

Underground installation   An extremely effective method of noise control is to bury the noise source underground. British Gas does this with pressure regulating stations and has by this means avoided complaints even where such installations have been sited near dwellings in residential or rural areas.

Partitions

In some situations partitions provide more suitable methods of noise control than enclosures. The design of a partition should be considered as part of a noise control programme when, for example, a large noisy room is to be subdivided to provide a quiet area, or when it is required to minimize the effect on a room of installing a noisy machine in an adjacent space.

If two rooms are adjacent with a common dividing wall, a sound source in one room will produce a sound field which will impinge on the wall, causing it to vibrate and radiate sound into the second room. The amount of sound radiated will depend upon the construction of the wall, particularly its mass, and the frequency of the sound. The harder and more reflecting the surface of the wall the less acoustic energy will enter it, and the more massive its structure the more difficult it is for the sound to set the wall into vibration and the less energy will be transmitted into the receiving room.

The sound insulation properties of a wall are usually expressed in terms of its sound reduction index or transmission loss. The sound reduction index, R, of a partition is related to its mass and the frequency of the transmitted sound by the mass law:

$$R = 20 \lg Mf - 47 \qquad \text{dB} \qquad (5.6)$$

where M is the surface mass of the partition, kg/m^2
   f is the frequency of the transmitted sound in Hz.

The mass law implies that the sound reduction index increases by 6 dB when the mass of the partition is doubled, but in practice the increase is nearer to 5 dB.

A single, average sound reduction index for a partition over the frequency range 100 Hz to 3.15 kHz can be calculated using the formula

$$R_{av} = 14.5 \lg M + 10 \qquad \text{dB} \qquad (5.7)$$

Equation 5.7 is accurate to within about 6 dB for the mean value of the sound reduction index for single-leaf, homogeneous partitions with no absorbent facing. It tends to overestimate the average sound reduction index for single complex partitions by between 2 and 10 dB, and to underestimate stud and double partitions.

Equations 5.6 and 5.7 are strictly applicable only to those frequencies below the critical. However, as the damping present in most building materials is relatively large they apply over most of the frequency range of interest.

On-site partitions will give less than the predicted sound reduction for all the reasons cited when discussing enclosures, but in addition there will always be some flanking transmission through the surrounding structures. Where a sound reduction index of 50 dB or more is required it is highly unlikely that a dividing wall alone will suffice because flanking transmission will be an important component of the sound heard in the receiving room.

It is also particularly important that there are no direct air paths linking the source and receiving rooms. The most common air paths around partitions are above suspended ceilings and through continuous convector casings and ventilation ducting. Special attention should be given to sealing such spaces and avoiding acoustic weak spots.

Prediction of performance   Where two rooms are divided by a partition the reverberant sound pressure level in the receiving room is given by

$$L_{pr} = L_{ps} - R + 10 \lg S - 10 \lg A_r \quad \text{dB} \tag{5.8}$$

where L_{pr} is the average reverberant sound pressure level in the receiving room, dB re $20 \mu Pa$
L_{ps} is the average reverberant sound pressure level in the source room, dB re $20 \mu Pa$
R is the sound reduction index of the partition, dB
S is the surface area of the partition, m²
A_r is the total acoustic absorption in the receiving room.

Where the sound power level, L_w, of the noise source is known the reverberant sound pressure level, L_{pr}, in the receiving room is predicted by

$$L_{pr} = L_w - R + 10 \lg S - 10 \lg(A_r \times A_s) + 6 \quad \text{dB} \tag{5.9}$$

where A_s is the total acoustic absorption in the source room.

Illustrative example   A source of sound power level 107 dB re 10^{-12} *W* in the 500 Hz octave band is to be placed in a plant room separated from an office by a 4×3m brick wall, 230mm thick, plastered on both sides. The total absorption in the plant room may be taken as 11.5 metric units, and in the office as 13.5 metric units.

What will be the likely 500 Hz octave band sound pressure level in the office due solely to the sound transmitted through the wall?

Such a wall has $R = 48$ dB in the 500 Hz octave band. The area of the partition is 12m².

The SPL in the office, L_{pr}, may be predicted using equation 5.9

$$L_{pr} = 107 - 48 + 10lg12 - 10lg(11.5 \times 13.5) + 6 \qquad \text{dB re } 20 \; \mu\text{Pa}$$

This gives a predicted reverberant SPL in the office, due to the new noise source in the plant room, of 54 dB re 20 μPa in the 500 Hz octave band. This ignores flanking transmission so the total level may be a little higher.

Composite partitions For the type of problem where the partition has to be constructed to include different elements, such as a door and window, consider the following illustrative example.

A 125mm-thick brick wall 5m long and 2.5m high forms the partition between two rooms. The wall is plastered on both sides and includes a door, size 2×0.75m. If the sound reduction indices in the 500 Hz octave band of wall and door are respectively 40 dB and 18 dB, find the overall sound reduction index of the composite partition.

To combine these values of R to give the average transmission loss for the entire partition the sound transmission coefficient, T, is introduced where

$$R = 10lg(1/T) \qquad \text{dB} \qquad (5.10)$$

or

$$T = \text{antilg}(-R/10) \qquad (5.10a)$$

The average sound transmision coefficient and loss for the partition are then found using

$$T_{ave} = \sum A_i T_i / \sum A_i \qquad (5.11)$$

and

$$R_{ave} = 10lg(1/T_{ave}) \qquad \text{dB} \qquad (5.12)$$

where A_i is the area of ith element of the partition, m²
T_i is the sound transmission coefficient of the ith element.

The areas and transmission coefficients for the partition are:

Item	A_i(m²)	R dB	T_i	$A_i T_i$
Wall	11.0	40	0.0001	0.0011
Door	1.5	18	0.01585	0.02377

$$T_{ave} = 0.02487/12.5 = 0.00199$$

and

$$R_{ave} = 27 \text{ dB}$$

Note that although the sound reduction index of the major element of the partition, the brickwork, is 40 dB, the poor insulation of the door reduces the overall performance to 27 dB. Areas of weak insulation reduce the overall performance of a partition and also cause relatively high sound levels in their immediate vicinity. The case history below illustrates the importance of ensuring that there are no areas of low sound reduction in acoustic partitions.

Double-leaf partitions   Experience has confirmed that double-leaf partitions can provide satisfactory insulation with much less mass than is needed with a single-leaf construction. The cavity should be wide enough to ensure that mass-spring-mass type resonances of the panels and the air in the cavity occur at frequencies too low to be troublesome. An estimate of the minimum separation required is

$$360(1/m_1 + 1/m_2)\text{mm} \tag{5.13}$$

where m_1 and m_2 are the superficial masses of the two leaves. Practical minimum separation between the panels should always be at least 100mm.

To avoid troublesome coincidence effects the two leaves should be of dissimilar surface densities, and as with enclosures the inclusion of a layer of absorbent material hung between the leaves improves the performance by up to 3 or 4 dB. See Figure 5.8 for a typical double-leaf construction.

Case history   A woodworking shop $28 \times 6 \times 3$m, had a typical noise level of 92 dB re 20 μPa in the 1 kHz octave band. It was required to create a noise refuge of volume $4 \times 6 \times 3$m, by constructing a partition to include a door and window across one end of the shop. The criterion for the shielded area was agreed as NR55 so a relatively massive partition was required.

The initial design, using materials to hand, was a double partition consisting of two leaves 100mm apart as shown in Figure 5.9. Each leaf consisted of a 50×100mm wooden frame faced with 15mm-thick plywood. The inside backs of the frame were filled with faced mineral wool held in place with wire mesh. The 2×1m door frame and the 1×2m window frame also acted as ties and the partition was fixed in non-setting mastic at the edges. The solid-core door was taken from stock and hung normally. The available material for the window was two sheets of 6mm glass, which were sealed into the frame, 200mm apart, using rubber compound.

The sound reduction indices, at 1 kHz, for the three sections of the partition are 49 dB, 29 dB and 56 dB, for frame, door and window respectively. In

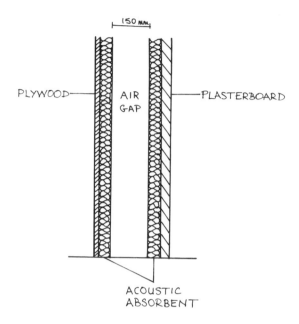

Figure 5.8   *Cross-section of a typical double-leaf partition*

addition there was an air gap around the door which may typically be assumed to be about 1 per cent of the door area with a sound reduction index of zero.

Item	Area A_i (m^2)	R (dB)	T_i	A_iT_i (m^2)
Frame	14	49	0.0000126	0.00176
Door	2	29	0.00126	0.00252
Window	2	56	0.0000025	0.000005
Air gap	0.02	0	1	0.02

$$T_{ave} = 0.00126, \text{ so } R_{ave} = 29 \text{ dB}$$

The average transmission loss of this substantial partition was calculated to be only 29 dB. The main reason was seen to be the air gap around the door. To improve performance the gap must be sealed with a form of draught-proofing sufficient to withstand industrial wear and tear, and the door fitted with 'pull-tight' handles. With these changes the average sound reduction index rises to 36 dB. Any further increase would require the acoustic performance of the door to be improved. This could be done by utilizing the thickness of the partition to have a double door arrangement.

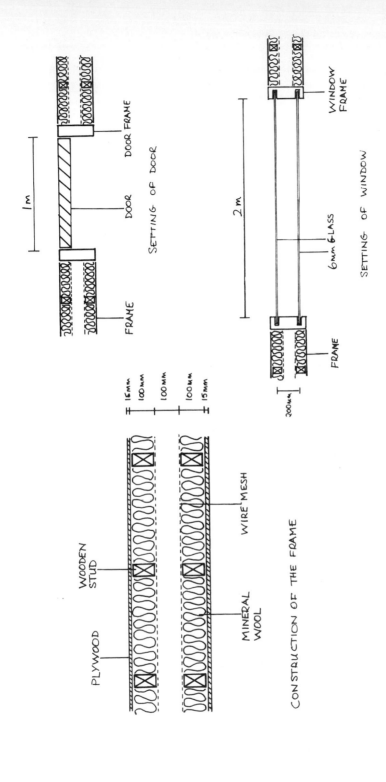

Figure 5.9 *Construction of partition in woodworking shop*

159

Using acoustic tiles the absorption in the receiving room was raised to 27 metric units and the insertion loss was then calculated as

$$IL = 36 - 10lg(18/27) = 38 \quad dB$$

In the real situation the absorption in the source room was also increased to reduce the noise level there.

Screens

In a situation where enclosures or partitions are not practicable a simple screen may produce worthwhile results. A screen is useful for producing a shadow effect in a given direction, determined by the relative geometry of source, screen and receiver. To be effective the screen must be close to either source or receiver, have a substantial sound reduction index and its smaller dimension must be at least several times the wavelength of the sound of the lowest frequency of interest.

Figure 5.10 specifies the various distances. Assuming the screen is positioned close to the source and the source-receiver separation is much greater than *h* the reduction in the direct sound pressure level may be taken as

$$10lg(0.06fh^2/r) \qquad dB \tag{5.14}$$

where   *f* is the frequency of the sound of interest
        h is as shown in Figure 5.10
        r is the distance from source to screen.

Unless the screen is relatively large, flanking transmission around it will limit the maximum attainable attenuation to about 20 dB. If the screen is some distance from either source or receiver very small reductions in the overall sound level will be experienced.

It is best to limit the use of screens to spaces where the screen height is at least one-third the height of the roof, though the higher the better. For reverberant spaces such screens will, on their own, provide some shielding up to about 5 to 8m away, but will have a noticeable effect for only 2 to

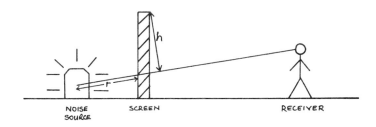

Figure 5.10   *Use of a screen to attenuate noise*

3m. The effectiveness of screens is increased by adding acoustic absorbent to the space, particularly the ceiling and those areas providing first reflections to the point in question.

Case history 1   A large compressor unit 2m high and comprising several noise sources caused complaints from residents at the site boundary. A long brick wall 5.5m high was built 3.5m from the compressor. Noise measurements made 15m from the compressor, 1m above the ground, before and after the introduction of the wall showed that the attenuations predicted using equation 5.14 corresponded closely with the actual sound reduction. The reasons for this were that the screen was sufficiently large for there to be little flanking transmission, and that because the compressor was outdoors there was no reverberant field to minimize the effect of the screen.

Case history 2   A large reverberant space contained a radial saw. A barrier 2.5m high (extending down to the floor) and 7.5m long was placed 1m away from the saw. The screen was 100mm thick, backed with 12mm plywood and filled with dense mineral wool held in place with galvanized chicken wire. The acoustic centre of the saw was taken as 1.5m above the floor. The sound spectra measured at 4m from the saw before and after insertion of the screen were as shown in Table 5.1

The screen, which is quite substantial, is effective only at the higher frequencies. If the noise had been predominantly low frequency the effect of the screen would have been insignificant. The efficiency of the screen was substantially reduced because of the high level of reverberant sound.

Noise reduction by acoustic absorption

A general and widely adopted approach to solving factory noise problems is the addition of acoustic absorbent to the room boundaries. It is a common

Table 5.1
Sound pressure levels, dB re 20 μPa, 4m from saw

	Mid-frequency octave band, Hz							
	63	125	250	500	1k	2k	4k	8k
Before screen	69	69	75	79	84	91	90	83
After screen	69	70	73	78	75	78	79	74
Measured attenuation	–	–	2	1	9	13	11	9
Predicted attenuation	–	–	10	13	16	19	22	25

experience that a sound source will appear louder in a space with hard interior surfaces (such as painted brick or smooth concrete) than it would outdoors. The reason is that sound is reflected back from the room surfaces thereby creating a reverberant field and increasing the sound intensity. Acoustic absorbent on the surface boundaries decreases the amount of reflected sound energy and so reduces the intensity of the reverberant field.

Application of acoustic absorbent is, superficially, the least cost effective of the methods discussed in this chapter. However, it has two decided advantages: the relative ease of selection and fitting, and its non-interference with the production process.

Prediction of performance

For typical spaces with sources evenly distributed this method will seldom reduce the reverberant field by more than 8 dB at any frequency or 6 dB(A) in total, and then only if the initial conditions are highly reverberant. The predicted drop in the reverberant SPL, on simple Sabine theory, is

$$10\lg(R_{c1}/R_{c2}) \qquad \text{dB} \qquad\qquad (5.15)$$

where R_{c1} and R_{c2} are respectively the room constants before and after the application of the absorbent.

Classical theory of room acoustics predicts that for a source of sound power, L_w, in an enclosed space the total sound pressure level at a point, L_p, will be given by

$$L_p = L_w + 10\lg(Q/4\pi r^2 + 4/R_c) \qquad \text{dB} \qquad\qquad (5.16)$$

where r is the distance from the source to the point in question
Q is the directivity of the source, which may generally be assumed to be 2 for machines on concrete floors
R_c is the room constant.

The uncritical use of the expression in equation 5.16 can lead to expensive errors. Nevertheless, it is extremely useful for demonstrating a severe limitation on the use of acoustic absorbent as a method of noise control. It is obvious that close to the source (r small) the direct field will dominate and the amount of acoustic absorbent in the space will have little effect upon the sound heard. Thus in those situations where the risk is due to noise from the worker's own machine, or others close by, it is unlikely that the addition of absorbent to the room surfaces would be of significant benefit. The most that could be expected would be a noise reduction of 2 dB(A).

Classical theory applies only to diffuse sound fields, which are not met in most practical situations. For large factory spaces where the shorter floor dimension is greater than four times the ceiling height, the effects of sound reflections from the side walls become negligible and the sound field approximates to that between two absorbing planes and decreases with distance from the source. Also there are usually a number of sources (the machines) and scattering elements (machines, cabinets, piles of materials, and so on) spread throughout the space which further reduce the applicability of the Sabine method.

The lack of a truly diffuse sound field in large factory spaces offers the opportunity for planning the space acoustically at the design stage. If the noisiest sources can be sited at one end of the factory, considerable attentuation may occur before the sound reaches the other end. For spaces over 50m long containing machines and with typical sound absorption coefficients for floor and ceiling, it should be possible to obtain a sound reduction of more than 30 dB between one end and the other.

For a space already in use, without repositioning of machines, the maximum likely additional decrease in sound level over the length of the space through an increase in absorption will be about 8 dB, though using the absorbers as additional scatterers might increase this to 12 dB if there was low absorption and few scattering centres originally.

Published data show that close to a single source, because of backscattering from nearby machines, the sound will be greater than that predicted by the Sabine method. Between about 1m to 2m and a distance equal to $0.2(A)^{0.5}$ (where A is the total internal acoustic absorption) the sound level will be within about 2 dB of that predicted by the Sabine method. For source-receiver separations greater than this the rate of decay of the sound field will be steeper, about 4 to 6 dB per doubling of distance (dB/dd) and for distances greater than say 40 to 50m the rate of decay will be 6 dB/dd or possibly more.

Figure 5.11 shows the decrease in sound pressure level with distance from a source for a factory space containing many scatterers. For rectangular spaces Lindqvist has produced *Design Curves for Estimating Sound Pressure Levels in Factories*,[8] but it has also been shown that roof shape has a considerable effect upon the sound propagation in large spaces.

Kuttruff[9] quotes a formula of 'good accuracy' for practical application. The sound intensity at a distance r from a source of sound power W is

$$I_r = \frac{(1-\alpha)W}{\pi ch^2} \cdot \frac{1}{(1+r^2/h^2)^{1.5}} + \frac{(1-\alpha)}{\alpha} \cdot \frac{b}{(b^2+r^2/h^2)^{1.5}}$$

where I_r is the sound intensity at a distance r from the source, W/m^2
h is the floor-ceiling height, m
c is the local velocity of sound, m/s
α is the mean absorption coefficient of floor and ceiling
the constant b is related to α as follows

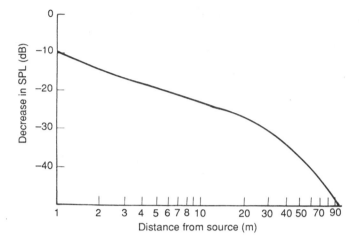

Figure 5.11 *A typical curve showing the drop in sound pressure level due to distance in a large rectangular factory space with low surface absorption* (After Lindqvist[8])

α	0.1	0.3	0.5	0.7	0.9
b	3.030	2.165	1.843	1.665	1.551

Before commencing any programme of retro-fit noise control by the addition of acoustic absorbent to the surfaces of a space containing many sources the situation must be assessed to determine whether the problem is caused by the reverberant field. If it is, and a likely maximum achievable reduction in sound level, without repositioning the machines, of 4 to 6 dB(A) is sufficient to justify the required expenditure, an increase in boundary absorption may be part of the overall solution.

The octave bands of interest should be determined because economically it is important to match the choice of absorber and its method of fixing to the problem at hand. Values of typical absorption coefficients for a large number of common materials and fixings are given in reference 7.

Methods of increasing absorption

The most common methods of introducing additional acoustic absorbent are:

1 Suspended absorbers. Suspending absorbers from the ceiling is doubly effective since when fixed in this manner they will also act as scatterers of sound. Usually the height of the absorbers will be determined by the need not to interfere with light fittings, fire sprinkler systems or crane operations. Commercially available products consist of semi-rigid compressed fibreglass or mineral wool slabs in various sizes and shapes. The lower the frequency

of the sound to be absorbed the thicker should be the absorber, with a practical upper limit of about 140mm. It should be noted that the absorption coefficients quoted in the manufacturers' literature are maximum figures which tend to decrease if the panels are placed closer together than 2m. For a given area of absorbent it is best to suspend the panels by their shorter side to give greater separation of the absorbers, while increasing their action as scatterers. Absorbers with a cylindrical form tend to increase low frequency absorption.

Where hygiene or aggressive environments demand, a thin protective layer may be placed over the absorber, though this will reduce performance at the higher frequencies, above about 4 kHz.

A mechanical repair shop used absorbent slabs of 1000 × 800 × 50mm thick at a density of one per square metre of plan area to reduce the reverberant noise level by 6 dB(A).

2 Acoustic sprays lack the action of scatterers and so are not generally as effective as suspended absorbers. For large floor areas it is best, if possible, to reduce the ceiling height before applying the material. Such spray-on materials can now be given a range of finishes to provide hygienic, waterproof or decorative properties. With the underside of a roof an additional consideration must be the thermal insulation properties of the material and consequent thermal energy savings.

3 Whereas the two methods above give broad-band absorption, resonators are selective (narrow-band) absorbers and are therefore useful in reducing noise levels where distinct tonal qualities are present. Selection and siting of resonators should be undertaken only with expert assistance.

Reduction of noise through good management

Workshop practices tend to have arisen through experience and tradition and thus can be changed successfully and permanently only if the workforce understands the reasons for the proposed changes. This method of noise reduction will be successful only if preceded or accompanied by an educational programme.

Change to workshop practices

Every factory suffers from noise due to poor maintenance or incorrect adjustment or operation of machines, for example leaking air lines, bearings in need of lubrication or balancing, faulty or missing silencers, noise control equipment or material abused, and so on. Such problems can usually be overcome easily and relatively cheaply. The benefits to be gained from attending to such details are invariably worthwhile and give a good return for the effort made. Indeed, attention to some of these details would normally bring economies of operation and a higher quality of work.

For example, a reduction of 3 dB(A) in the noise from a guillotine was obtained by changing the loading procedure to reduce clatter of the metal being cut on the machine surface.[5] In another factory, moving a skip closer to a worker and decreasing the height of fall of the cut pieces of metal simultaneously lowered peak noise levels and reduced damage to components.

Change in work processes

In some cases where it is impracticable to reduce noise by other means, substitution of a different method of operation often affords satisfactory noise reductions. In one example crimping rather than riveting reduced noise levels by over 10 dB(A).

Change in work procedures

Where individual workers undertake a range of jobs it may be possible to significantly reduce the noise exposure of workers by a programme of job rotation. Work can be rearranged to allow part to be carried out in a quiet place so that the time spent in the noisier areas will be divided between those on the rotation scheme. Additionally a noise refuge should be provided near the place of work.

Conclusions

The increased public awareness of noise and its potential for physiological damage and the preparedness of government bodies to legislate on these issues have made noise control an important topic for the industrial community. Fortunately considerable expertise is available, capable of balancing the various concerns and pressures to obtain a workable solution in the majority of cases.

Where none of the above means for controlling noise in the workplace reduces the noise exposure to an acceptable level, employed persons should be supplied with effective hearing protection individually. Provided that adequate protection is given it is preferable for the user to be allowed a personal choice amongst the different protectors available. This topic is discussed in detail in Chapter 6.

References

1   Council Directive 86/188/EEC of 12 May 1986, *Official Journal of the European Communities*, No. L137.
2   Department of Employment (1972) *Code of Practice for Reducing the Exposure of Employed Persons to Noise*, HMSO, London.
3   Chartered Institution of Building Services Engineers (1973) *Guide*, Section B12, Noise Control.

4   American Society of Heating, Refrigeration and Airconditioning Engineers (1984) *Handbook*, Systems Volume, Chap. 32.
5   Health and Safety Executive (1983) *100 Practical Applications of Noise Reduction Methods*, HMSO, London.
6   Bazley, E. (1966) *The Airborne Sound Insulation of Partitions*, National Physical Laboratory, HMSO, London.
7   US Department of Health, Education and Welfare (1980), *Compendium of Materials for Noise Control*, Publication No. 80–116, May.
8   Lindqvist, E. (1982) *Design Curves for Estimating Sound Pressure Levels in Factories*, Swedish Council for Building Research, Document D15.
9   Kuttruff, H. (1985) *Sound Propagation in Working Environments*, Proc. 5th FASE Symposium, Thessaloniki, Greece.

Chapter 6

HEARING CONSERVATION PROGRAMMES

Roger Wills, Senior Chief Audiology Technician, Royal
Berkshire Hospital, Reading

Deafness or hearing loss caused by noise is certainly not a new problem.
Gun crews during the Napoleonic wars became deaf, and ever since the Industrial Revolution it has been recognized that certain occupations carried the
risk of hearing loss. Terms such as 'weavers' deafness' and 'boilermakers'
deafness' became part of the language in medical circles, and among the people
who lived with these industries.

Noise-induced hearing loss has its main effect in reducing the sensitivity
for hearing high frequency sounds. Severe high frequency hearing loss leaves
speech audible but not intelligible, music becomes unrecognizable, and understanding a single speaker in groups socially or at meetings becomes impossible.
Hearing loss in all its various forms is an underrated disability, but this is
especially true of noise-induced hearing loss. These hearing changes cannot
be cured medically or surgically, and even the use of hearing aids may not
give good results since the damaged hearing usually has poor tolerance of
amplification, and considerable distortion of sounds occurs.

Criteria

There are very few working environments where the potentially hazardous
noise level is steady and continuous throughout the working period. For this
reason it has become common practice to use the concept of *noise dose* to
describe the extent to which workers exposed to noise may be at risk of
hearing damage. A worker's noise dose is calculated or measured so that
despite the fact that the actual noise level is fluctuating, the amount of acoustic

energy affecting the worker can be described using a single figure in dB(A). This equivalent continuous sound level normalized to eight hours (L_{Aeq} 8 hour), is the notional continuous sound level which would in eight hours give the same dose of acoustic energy as the actual fluctuating sound level. Commercially available personal noise dose meters, or integrating sound level meters will give an eight-hour L_{Aeq} reading. It is important that the measurement period is long enough to be truly representative of the real noise exposure situation. For complete accuracy the whole eight-hour period must be measured. This concept of eight-hour L_{Aeq} or *Daily acoustic immission* (DAI L_{Aeq} dB(A)), greatly simplifies the problem of understanding and dealing with the hazard of noise in the workplace.

Following the 1972 *Code of Practice for the Reduction of the Exposure of Employed Persons to Noise*,[1] the UK has adopted a noise dose limitation of 90 dB(A) L_{Aeq} as the current working standard. There would seem to be a trend towards a lower level in many countries, and it is not unusual for companies independently to adopt an 85 dB(A) L_{Aeq} as their internal standard even in the UK.

The recent consultative document *Prevention of Damage to Hearing from Noise at Work*,[2] issued by the UK Health and Safety Commission (HSC), seeks comment on draft proposals to implement the 1986 European Community Directive 86/188/EC. This directive is to be implemented in most member states by 1990, and will require some changes in the existing UK legislation.

The HSC draft proposals suggest three 'action levels' at which employers and employees will be required to take certain steps with regard to hearing conservation measures. The first action level is a personal noise exposure of 85 dB(A) L_{Aeq}. This is followed by 90 dB(A) L_{Aeq} as the second action level and a third or 'peak' action level at 200 pascals (140 dB re 20 μPa). The scope of these new proposals is greater than the 1972 code and the later HSC document *Framing Noise Legislation*, and it is to be hoped that implementation by 1990 proves possible and acceptable to all concerned.

In the past even the existence of draft proposals such as these has influenced health and safety strategies and possibly legal opinion, about the current state of commonly available knowledge in this area. The time scale for changes in legislation may mean that many employers will in practice have begun to act in accordance with this new thinking long before they are obliged to do so.

This growth of legislation and guidance on industrial safety matters has led to a greater awareness that hearing conservation is as deserving of attention as the many other more commonly discussed workplace hazards. This change in attitude has been encouraged by legal judgments which have clearly recognized that an employer can no longer claim ignorance of the possible physiological effects of high noise levels, and so must take steps to protect the workforce.

A simple method for estimating L_{Aeq} is given in the nomogram in Figure 6.1, which is taken from the 1981 consultative document on noise.

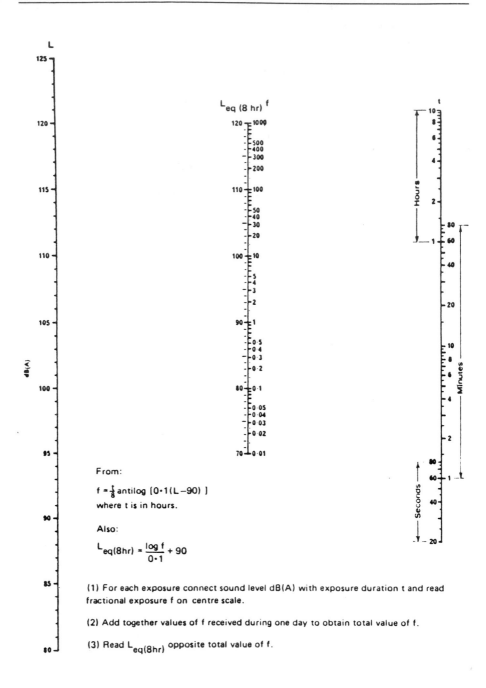

From:

$$f = \frac{t}{8}\,\text{antilog}\,[0{\cdot}1(L-90)\,]$$
where t is in hours.

Also:

$$L_{eq(8hr)} = \frac{\log f}{0{\cdot}1} + 90$$

(1) For each exposure connect sound level dB(A) with exposure duration t and read fractional exposure f on centre scale.

(2) Add together values of f received during one day to obtain total value of f.

(3) Read $L_{eq(8hr)}$ opposite total value of f.

Figure 6.1   *Nomogram for calculation of equivalent continuous sound level, L_{Aeq}* (Reproduced with the permission of the Controller of Her Majesty's Stationery Office from the 1981 Consultative Document on Noise.)

The nomogram is used by means of the following steps.

1 Ignore exposures of less than 85 dB(A).
2 Draw a straight line to connect the measured sound level in dB(A) on the '*L*' scale, with the exposure duration on the 't' scale.
3 Read the value of '*f*' on the centre scale. Do this for each exposure level.
4 Add all the '*f*' values received during the working period, and read the L_{Aeq} value opposite to the total '*f*' value.

This method is useful only if the sound levels are steady over particular periods but vary throughout the day. The increased sophistication of modern acoustic measuring equipment has now meant that the eight-hour L_{Aeq} can be measured directly and relatively cheaply.

Hearing conservation programmes

Surveys

The flowchart shown in Figure 6.2 illustrates the design of a typical approach to hearing conservation. The first step in the programme will be the measurement of noise levels throughout the working area. For more detailed guidance on noise measurement see Chapter 2. A noise survey should ideally include at least the following elements.

1 Identification of the peak noise exposure levels for all employees, in dB(A) and by octave band analysis.
2 Noise dose (DAI L_{Aeq} dB(A)) information for all workers exposed to peak levels beyond 80 dB(A).
3 A noise level map of the whole work site, showing peak and L_{Aeq} levels.

It may be necessary to use tape recordings of the noise if the nature of the exposure is complex. This will permit statistical time distribution analysis to show the proportion of time that a given sound level is exceeded.

Careful consideration should be given to the choice of instrumentation for noise measurement. Some personal noise dose meters show a reading in accumulated L_{Aeq}, whilst others show a figure or binary code representing the percentage of the maximum permissible noise dose. The design of some dose meters permits confidentiality of the final readings, while on others it is visible to the worker. Dose meters are open to abuse by, for example, removing them to a quieter or noisier place for the test period, or by shouting into the microphone. Integrating sound level meters or personal dose meters do not all have the same design standard in terms of operating technique and dynamic range. In Europe a 3 dB value to represent a doubling of energy has been accepted, whereas in the USA 5 dB is used. An illustration of the

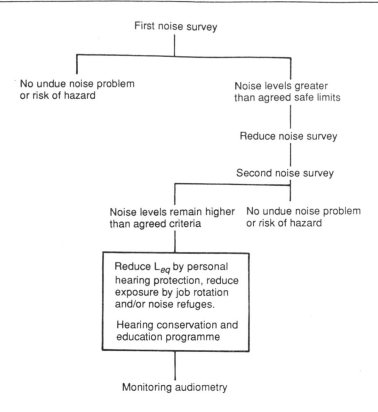

First noise survey

No undue noise problem
or risk of hazard

Noise levels greater
than agreed safe limits

Reduce noise survey

Second noise survey

Noise levels remain higher
than agreed criteria

No undue noise problem
or risk of hazard

Reduce L$_{eq}$ by personal
hearing protection, reduce
exposure by job rotation
and/or noise refuges.

Hearing conservation and
education programme

Monitoring audiometry

Figure 6.2   *Flowchart for hearing conservation programme*

UK 90 dB(A) L_{Aeq} energy doubling standard is shown in Table 6.1.

For the purposes of this chapter it is assumed that all practicable engineering noise control has been achieved before this working noise hazard survey is carried out. The survey will identify the working noise exposure for the whole workforce. Consideration in line with policy or national legislation can then be made as to whether any hazard to hearing exists. In addition, it will be possible to define the noise climate around the work site. This will allow the creation of known noise zones, where, for instance, entrance is permitted only with the use of appropriate hearing protection. Entrances to these areas should be clearly signed, perhaps with colour-coded signs to make the zoning and type of hearing protection obvious to all.

There may be the need to consider other strategies to avoid some noise exposure for particular workers. In circumstances where machines cannot reasonably be enclosed, perhaps the workers can be given a sound-treated noise refuge in the noisy area (see Chapter 5). Process monitoring controls can be arranged within the enclosure, so that working time in the higher noise levels outside is kept to a minimum. An addition or alternative to this

Table 6.1

Showing equal noise dose values based on the UK 90 dB(A) 8-hour L_{eq}

Duration of noise exposure	Noise level
8 hours	90 dB(A)
4 hours	93 dB(A)
2 hours	96 dB(A)
1 hour	99 dB(A)
30 minutes	102 dB(A)
15 minutes	105 dB(A)
7.5 minutes	108 dB(A)
3.75 minutes	111 dB(A)

possibility is to look carefully at shift patterns and job rotation during shifts, as a means of controlling total noise exposure time.

Hearing protection

Those workers whose noise exposure exceeds the L_{Aeq} criteria chosen will need to use hearing protection so that their effective L_{Aeq} can be within safe limits. There is a very large range of hearing protection devices now available. Examples of the main classes of protectors are listed and described below.

Single-use, disposable standard-sized earplugs, in rubber, soft plastic (V–51R plugs), glassdown, or wax material

This type of earplug can cause problems of discomfort in use. Ear canals vary considerably in shape and size, and one person's left and right ears may not even be the same size. There is a danger therefore that plugs may be chosen for comfort rather than adequacy of fit. In practice the so-called standard-sized plugs are seldom a satisfactory fit, but the glassdown preformed plugs, which will conform without discomfort to the ear shape, can be quite satisfactory. Unfortunately the material resembles cotton wool, which gives the impression that it has the same attentuation. In fact cotton wool has almost no attenuation at all since the fibres are far too large to give adequate density when compressed. Some of the standard earplugs are available joined together with a thin spring steel headband. The spring tension applied by the headband can greatly help the otherwise widely variable attentuation of these standard earplugs. It should be noted that the steel headband would be detected by magnetic screening devices used in some industries to find foreign bodies in the product.

Special expanding foam plastic universal fit earplugs (EAR type)

Earplugs made from this special material, which looks like the foam plastic used in upholstery, have the unique property that when compressed the plug takes some twenty to thirty seconds to recover its shape. There is sufficient time for it to be rolled into a small cylindrical shape between finger and thumb, inserted into the ear and held briefly while it expands to accommodate the ear canal shape. The elastic memory of the material exerts a continuing force to retain the fit and attentuation throughout the working day. These plugs give good performance in comfort and attentuation, can be washed and re-used a number of times, and provide a viable alternative to earmuffs.

Custom-made individually moulded earplugs (personal semi-insert type)

For maximum comfort and security in the ear these individually moulded plugs will always tend to perform best. Ear impressions must be taken of every person who is to be fitted with these plugs, using a silicon rubber impression material. Millions of hearing aid users have earpieces made using the same techniques, and are able to wear them without discomfort all their waking hours. The impression can be taken only by a trained person, and the bespoke nature of these plugs is an undoubted drawback, but they do have some useful applications, and can give good attentuation.

Headphone-style earmuffs

These are available in many designs and cover a wide range of degrees of attentuation. Often these protectors can be supplied in a form which allows for fixing to hard hats or safety helmets, although this may have implications for the original design of the hat. Any accident involving the depression of the hat towards the skull could force the earmuff shells over the ears causing additional injury. The attenuation of normal earmuffs depends on the softness of the seals around the ears, and the headband tension which must not be altered by bending or stretching.

Earmuffs which incorporate the electronics of 'anti-noise' or active noise control are now available to give increased attenuation for low frequency sound. This rapidly developing technology will have a most important place in the future of hearing protection, since mechanical means of attentuating sound have always given poor results for high levels of low frequency noise. This topic is discussed in Chapter 8.

It is possible to use a combination of earplugs and earmuffs together, but the attenuation will not be additive. There will probably be an extra 5 dB over the earmuffs used alone. In noise levels over 120 dB(A) much sound will be passing through the bones of the skull regardless of ear protection. Noise protection helmets should be worn in these circumstances, but no part of the body should be exposed to levels beyond 150 dB (re 20 μPa).[1]

Performance

An important measure associated with attenuation is the amount by which actual attenuation will vary from one individual to another. Manufacturers' figures will quote the mean attenuation and the variability or standard deviation. Table 6.2 shows the performance of a range of hearing protection with the standard deviation shown in brackets. These figures were achieved using the British Standard test method (BS 5108).[3]

Protectors are made which aim to impede normal communication as little as possible. Some of these types of hearing protectors are available with frequency-selective attenuation, but since this will usually provide minimal attenuation for low frequencies, it is not suitable for most industrial use. In industrial settings the highest noise levels are often of a low frequency nature. Similarly, those devices which provide amplitude-sensitive attentuation (ear-valve type), useful against gunfire and explosive noise, are not usually suitable for the industrial environment.

Some hearing protection can be fitted with a means of receiving music or messages through an induction loop system. There will be a limiter built in so that the music level cannot itself be hazardous. An example of a more sophisticated version of this arrangement is the tethered two-way communication combined with hearing protection, used by ground staff when working with aircrew as aircraft are being moved around airport hardstanding.

Choice of hearing protection

The choice of the most appropriate hearing protection will depend on a number of factors, ranging from the ergonomic to the technical. The protection should be on personal issue and spares and replacements should be easily available. Ideally there should be more than one class and type of protection available within the technical constraints, so that workers can be involved in the choice for their own use. A small incidence of workers with a history of ear disorders of various kinds will be found, which may prevent the use of some types of protection. Medical advice should be sought in these cases.

The technical and practical considerations for choice of suitable protection are as follows.

1 Adequate sound attenuation.
2 Compatibility with other safety or hygiene equipment used by the worker.
3 Compatibility with hair style and items worn on the head; the workers' spectacles, ear rings, hair pieces, hearing aids, and so on.
4 Communication and general safety needs.
5 Comfort and hygiene.
6 Cost.

Adequate sound attentuation

The first criterion, adequate sound attenuation, can be calculated by the use

Table 6.2

Mean values of attenuation for seven hearing protection devices (after Martin)[4] evaluated and presented in accordance with BS 5108.[3] (Some of these items may be no longer available commercially.)

Protector	Test centre frequency (Hz)									
	63	125	250	500	1k	2k	3.15k	4k	6.3k	8k
Bilsom propp (glassdown)	4 (3)	6 (4)	8 (5)	11 (5)	15 (4)	19 (5)	26 (5)	26 (5)	33 (6)	35 (5)
Norton sonic ear valve 143.4D	1 (4)	1 (3)	0 (4)	0 (5)	9 (4)	16 (5)	19 (4)	15 (4)	20 (7)	14 (6)
Mine safety appliances V51–R	20 (6)	20 (7)	19 (7)	19 (7)	22 (6)	27 (5)	32 (5)	29 (6)	29 (12)	30 (10)
Instaform protectears (Personal silicon earplugs)	19 (8)	18 (8)	17 (7)	23 (7)	21 (6)	29 (6)	39 (7)	42 (7)	35 (10)	39 (9)
EAR foam plastic plugs	25 (7)	26 (8)	27 (7)	29 (7)	30 (6)	33 (5)	43 (5)	44 (5)	45 (5)	44 (6)
Amplivox supamuff (foam seal)	9 (4)	7 (4)	8 (4)	16 (5)	26 (6)	27 (5)	36 (5)	34 (5)	29 (5)	30 (7)
Amplivox sonoguard (liquid seal)	17 (6)	19 (5)	24 (4)	32 (4)	41 (5)	38 (5)	43 (4)	42 (6)	29 (5)	32 (6)

of data from the noise survey and the hearing protection manufacturers, in conjunction with national standards. This process will begin with the definition of the 'assumed protection' given by the hearing protector. The degree of protection which most people will obtain is equivalent to the mean attenuation value minus the standard deviation. Table 6.3 shows an example of this.

The assumed protection is next compared with the octave band noise levels from the noise survey, to arrive at the estimated sound levels at the user's ears under the ear protection. Table 6.4 shows this stage in the calculation.

These estimated octave band sound levels at the user's ear are then 'A' weighted, and converted into arbitrary intensity units 'I' using Table 6.5. When these intensity units have been added together the resulting figure is converted back into a single figure in dB(A) using Table 6.5 again. These stages are illustrated in Table 6.6.

This method may be used for all the types of hearing protection being considered, so that adequate alternatives can be provided for each location.

Care should be taken to avoid the selection of protectors which give too much attenuation. The only effect of reducing noise levels considerably beyond that necessary for safety will be to risk the rejection of the hearing protector

Table 6.3

The derivation of the assumed protection for a hearing protector

	Frequency (Hz)						
	125	250	500	1 000	2 000	4 000	8 000
Mean attentuation dB	15	20	34	36	40	44	42
Std deviation dB	5	6	5	5	6	7	8
Assumed protection dB	10	14	29	31	34	37	34

Table 6.4

The deduction of the assumed protection from the noise, to arrive at the SPL at the user's ear

Octave band centre frequency (Hz)	Octave band analysis of noise (dB re 20 μPa)	Assumed protection, from Table 6.3 (dB)	Assumed SPL at user's ear (dB re 20 μPa)
125	93	10	83
250	95	14	81
500	100	29	71
1 000	103	31	72
2 000	107	34	73
4 000	100	37	63
8 000	88	34	54

Table 6.5

The values of the arbitrary intensity units 'I' corresponding to SPLs from 60 dB to 119 dB

dB	0	1	2	3	4	5	6	7	8	9
60	0.0010	0.0013	0.0016	0.0020	0.0025	0.0032	0.0040	0.0050	0.0063	0.0079
70	0.0100	0.0130	0.0160	0.0200	0.0250	0.0320	0.0400	0.0500	0.0630	0.0790
80	0.1000	0.1260	0.1580	0.2000	0.2510	0.3160	0.3980	0.5010	0.6310	0.7940
90	1.0000	1.2600	1.5800	2.0000	2.5100	3.1600	3.9800	5.0100	6.3100	7.9400
100	10.0000	12.6000	15.8000	20.0000	25.1000	31.6000	39.8000	50.1000	63.1000	79.4000
110	100.0000	126.0000	158.0000	200.0000	251.0000	316.0000	398.0000	501.0000	631.0000	794.0000

Table 6.6

The conversion of octave band SPL at the user's ear to dB(A). The 'I' value of 0.0708 corresponds to 79 dB(A) to the nearest dB.

Octave band centre frequency (Hz)	Assumed SPL at user's ear, from Table 6.4 (dB)	'A' weighting correction (dB)	'A' corrected octave band level (dB)	'I' intensity units, from Table 6.5
125	83	− 16	67	0.005
250	81	− 9	72	0.016
500	71	− 3	68	0.0063
1 000	72	0	72	0.016
2 000	73	+ 1	74	0.025
4 000	63	+ 1	64	0.0025
8 000	54	− 1	53	−

Total value of 'I' = 0.0708

by the worker. Unreasonably high attenuation will interfere with communication, isolate the worker from the environment and may give rise to its own safety problems.

Compatibility with other safety and hygiene equipment

The compatibility with other safety equipment can present some problems. Earmuffs may be difficult to wear under a welding helmet, and earplugs may prove more acceptable, assuming that the attenuation has been shown to be adequate. Earmuffs are often made in bright colours, which can be a useful feature, allowing them to be seen in use from some distance away. In some industries, such as food processing, there may be great concern regarding the possible contamination of the product by lost earplugs. It may be more satisfactory to use tethered earplugs or headband earplugs in these cases. Earmuffs may not be an easy alternative since the common warm environments in these industries may lead to discomfort.

Compatibility with other head-worn items

Spectacles of all types may interfere with the fit and consequently the performance of earmuffs. This should be borne in mind even if workers are not objecting to the use of earmuffs with spectacles. If the spectacles are used as eye protection an alternative type of eye protection or earplugs may provide a solution. Ear rings, hair styles and headwear used by some religious groups can have the same effects, interfering with either comfort or the fit of earmuffs.

Communication and general safety needs

The implications of the use of hearing protection on the ability of the worker both to communicate with colleagues and to receive messages are of great importance. In many industries it is inevitable that some workers will be found to have a slight hearing loss already. Amongst any large group of adults this will be true, but especially amongst older workers with many years' background of military and/or occupational noise exposure. When hearing protection is used over an existing slight hearing loss the problems of acclimatizing to its use may be greater than average. This issue should be considered before a hearing conservation programme begins, since failure to do so will often be a factor in cases of poor co-operation with the programme. If workers are used to music in the workplace, amongst all the other noise, they may well object to its no longer being available. Similarly, it may be harder even for people with good hearing to hold a conversation with a colleague in the early stages of becoming accustomed to hearing protection. Some of these problems can be prevented by effective programmes of safety education. Consideration should also be given to the many advantages of earmuff hearing protection with built-in music and paging pick-up, by means of an inductive loop system. This might help the acceptance of the protection since the music reception will be of far better quality than that which the workers would previously have experienced. It will also be possible to use the paging system to make staff announcements and as an additional means of emergency warning system.

Working areas where other safety hazards exist may need careful replanning when hearing protection is introduced. All auditory warning signals such as bells for the movement of overhead cranes, or sound signals for process monitoring may need to be amended. Light signals may now prove more effective, and reorientation of work stations to bring visual signals into the workers' field of vision may be necessary. Those machines or processes that workers perceive as monitored by them by hearing, may need careful reassessment. The workers' perception may be correct, and new warning devices, may be justified to protect the machine or process. Sometimes the perception is not valid, and the comment may represent token or serious resistance to the use of hearing protection. Safety education programmes as a part of the hearing conservation programme will usually overcome these problems.

There may be a need for marked traffic lanes for vehicles such as forklift trucks, to reduce the likelihood of accidents. This type of obvious planning for the hearing conservation programme will have a beneficial effect on the workers' acceptance of the outcome. Similarly, management and visitors must be consistent in their use of hearing protection in the appropriate noise zones of the site.

Comfort and hygiene

The policy of making hearing protection available by personal issue will usually promote better utilization, as workers do not want this type of safety equipment to have been previously used by others. There should be a clear policy of regular replacement of disposable parts of the equipment, such as earmuff ear seals. Disposable earplugs and spare parts for other ear protection should be readily available in locker rooms and near to the workplace. Workers must also be permitted to change their minds, in the light of experience about the type of protection they wish to use.

The education programme which should accompany the introduction of hearing protection should stress the importance of cleanliness in the use of protectors. Earplugs must be inserted by clean hands before starting work, and not removed at any time except after washing the hands again, especially if the workplace is not itself clean. This routine becomes more important when substances such as oil or solvents will be on the workers' hands at the end of their working periods. Under no circumstances should earplugs be re-used by other people.

Cost of hearing protection

The total cost of a hearing conservation programme cannot easily be estimated in advance. The various elements of noise surveys, engineering noise control, repeat noise survey and ongoing conservation programme, will reveal their costs as the whole scheme proceeds. The noise survey may be carried out by an outside agency, at a quoted cost, but engineering noise control costs can be very considerable or modest depending on many factors. Hearing protection costs will be predictable once the scale of the remaining problem has been assessed. Certainly it is possible to restrict costs by reducing choice of protection devices, or by shift workers sharing safety equipment, but these are false and counter-productive economies if utilization is to be maximized. Engineering noise control can be costly, but represents a one-off cost that can permanently improve the quality of the working environment. Hearing protection is a continuous cost, and to some extent should be seen as a means of controlling the hazard while finding better and more permanent solutions. A company noise specification should be written into all future contracts for new plant and machinery to avoid the acquisition of more noise problems. Only by this means can proper control be exercised over the future noise climate.

Administrative costs will be associated with the programme: ordering stock, stores, education and training, hire of films, talks, posters and signs. Over a period of time it will become possible to compare the hearing protection costs with those of engineering noise control.

While considering the cost-effectiveness of noise control and hearing conservation, it may be helpful to draw attention to the past reports of some researchers into noise effects. Abey-Wickrama *et al.*[5] Broadbent,[6,7], Fox,[8] Hockley,[9,10] and Weston and Adams[11] are among many who have sought to show the effects of noise on general physiology and performance of tasks.

It has become generally accepted that while simple tasks are not impaired by the presence of noise, indeed they may even be improved by low-level noise, more complex tasks are performed progressively less well with increasing noise levels. A study of this previous work would support the probability that noise control and hearing protection can lead to increased efficiency, and consequently to the implied financial benefits.

Educational needs in the hearing conservation programme

At all stages in the planning and implementation of the programme it will be found very beneficial to pay attention to spreading information about the scheme to all grades of the workforce. Workers and management are likely to respond more positively towards a programme which they have had an opportunity to understand. There will naturally be mistrust and anxiety over any programme introduced without adequate consultation. The safety representatives of the labour organizations should be involved with the health and safety team and employers at all stages.

Although the concepts and the physics behind hearing, sound, hearing damage and noise control are complex, this need not prevent the presentation of information on the background to noise control and hearing conservation. A guiding factor in talks to people in an industrial setting should be to aim the presentations at an adult further education level. The material used must be of a very professional standard, and the educational sessions should be designed to cover specific topic areas concisely. Where the sessions are held in work time it is crucial that good liaison be maintained between the safety trainers and the departmental managers. Sessions should run to an exact time-table, starting and finishing on time every time. The nature of the education and safety training should not be seen as tied to a classroom or lecture-type format. This can be the most difficult thing to do well, particularly for safety staff who have no formal training in adult further education techniques. The education programme should have clear objectives, and each should be considered for presentation in a variety of forms.

The objective of maximizing the use of hearing protection, for instance, could be partly covered by means of films or videos from the large range available through suppliers of safety equipment, or from government agencies. Do make sure that the material in the film is relevant to the industry in question. The attention it gets and the informaton it succeeds in imparting, will have a great deal to do with the audience's perception of its relevance to them. Even an American accent can interfere with the credibility of the material when heard by an English audience, and vice versa. The film could be backed up by a workplace poster campaign, with regular changes of material to maintain interest. In addition the consistent use of hearing protection by supervisory and management staff must be guaranteed. Notices around the work site should clearly identify areas where protection must be worn. There should also be a regular check on the state of individual worker's protectors.

This can be achieved simply by giving one worker in each department the responsibility to go around offering spares and replacements at regular and predictable intervals.

Safety induction courses for new workers must be devised to include the needs of the hearing conservation programme. These new people will provide the best long-term data on the value of the programme, especially those who are young and not previously exposed to noise. The young are also the most vulnerable to hearing loss, since their hearing is at its most sensitive and easily damaged. If the young can be encouraged to understand the need for noise control and the use of hearing protection, they will be more likely to continue with its use and will derive the greatest benefit

The health and safety team should keep comprehensive and accurate records of all aspects of the programme. These must include noise survey data, evidence of attendance at training sessions, the issue of protection devices, poster campaign schedules and employee records. These records are an important medico-legal safeguard for all parties concerned.

Audiometry in hearing conservation programmes

The general safety policy for those work environments which include significant noise exposure will need to involve some overall safeguard to ensure the effectiveness of the hearing conservation programme. This safeguard can best be provided by the use of a hearing test, known as audiometry, for pre-employment confirmation of hearing status, and regular monitoring of the hearing of employees. Audiometry is the measurement of hearing sensitivity for a range of pure tone sounds, usually between 250 and 8000 Hz. The tests are carried out using headphones to allow the test sounds to be played separately to either ear. The subject indicates when the sound is heard by means of a press button which is held down until the test sound becomes too faint to be heard, when it is then released. The audiometer will then again increase the intensity of the sound until the button is pressed once more. Each pure tone is played by the audiometer for about thirty seconds, which is long enough to establish the minimum audible threshold, before switching automatically to the next frequency. When the final frequency has been tested on the first ear the audiometer will switch the sound to the opposite earphone. These self-recording audiometers designed for an occupational health role have a built-in plotter which draws the result as the test progresses. The finished audiogram will appear as shown in Figure 6.3, which shows a typical slight noise-induced hearing loss. The top horizontal line at 0 dB represents clinical normal hearing. On the audiogram it will be marked in dBHL (hearing level) to denote that it is derived from a series of corrections to produce the straight line at zero. Human hearing is less sensitive at either extreme of the audible frequency range, especially at low frequency, and this fact is taken into account to obtain a straight line for 0 dBHL. This is an average of the hearing of otologically normal young people, and consequently the further down the chart a particular audiogram comes the greater is the hearing loss involved. Good hearing is especially necessary in the mid to high frequencies for the clear discrimination of speech. Throughout life the sensitivity for hearing higher frequencies deteriorates slightly for the majority of people as a natural effect of ageing. Less sophisticated manual audiometers can be used for testing small numbers of people, but the tests will take longer and introduce an extra variable into the result. The operator of a manual audiometer must make a judgement of the level of the subject's hearing threshold, and this judgement requires greater skill and vigilance than the operation of self-recording audiometers.

The dip in the audiogram of a noise-induced hearing loss is a very common finding. It will usually appear at around 4000 Hz, and its cause – although still not fully understood – seems to relate to the physiology of the hearing mechanism, and its behavior in response to very high energy inputs. There are other types of hearing disorder which can cause a dip in an audiogram at or near 4000 Hz, but noise is much the most common cause of this finding. In terms of hearing change which would be noticeable to the worker, the dip shown in Figure 6.3 would be of no consequence. However, it does provide a very sensitive way of detecting the hearing changes most commonly associated with noise exposure, before significant damage can occur. Once the dip extends down to about 45 dBHL, it is likely that some social hearing disadvantage will begin to be apparent. This will typically cause problems in the discrimination of some speech sounds, especially in the presence of competing noise. More severe hearing loss in this high tone area will cause progressively greater difficulties. Figure 6.4 shows the appearance on an audiogram of a more marked noise-induced hearing loss, showing how the problem typically progresses over the years of noise exposure.

The introduction of an audiogram into the pre-employment medical will

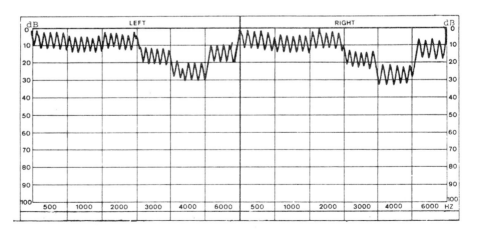

Figure 6.3   *Audiogram – slight noise-induced hearing loss*

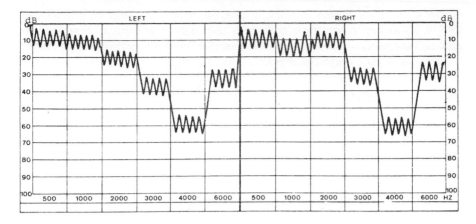

Figure 6.4   *Audiogram – more marked noise-induced hearing loss*

safeguard both parties. The prospective employee will have evidence of his or her hearing status before any noise exposure with that employer, and the employer can see both that the hearing is adequate for the job involved, and that any hearing loss found at this stage cannot be held to be the firm's responsibility. The discovery of some hearing loss need not preclude the employment of the candidate, assuming that it is not likely to interfere with the ability to do the job involved correctly and safely.

Test conditions

Audiometry will be of value only if it is carried out in an appropriate acoustic environment. Shipton and Robinson[12] have defined the maximum permissible ambient noise levels for accurate audiometry. Using the conventional headphones accurate results are possible up to an ambient noise level of about 30 dB(A). If this proves difficult to achieve the headphones can be supplied with sound-attentuating enclosures, allowing testing in up to about 50 dB(A) of ambient noise (see Figure 6.5). If the ambient noise is higher, small sound-treated booths are available, designed especially for audiometry. The test candidate sits inside the booth with the headphones and signal button, with the audiometer and operator outside visible through the booth windows (see Figure 6.6). Spurious and variable results will be obtained unless this detail of the audiometric environment is properly arranged. The planning for audiometry should also take into account the need to avoid testing too soon after actual noise exposure. The ear suffers temporary reduction of hearing when exposed to high levels of noise, and this temporary threshold shift (TTS) will make the audiogram show a greater hearing loss than the subject actually has. If possible audiometry should be done after a break of two days from noise in order to avoid TTS, but this is often impracticable. Two compromises

Figure 6.5   *Typical headphones for audiometric tests*
(Reproduced by permission of Amplivox Ltd.)

can be considered: first, to allow at least sixteen hours free of noise prior
to the test and, second, to test during the working shift provided that appropriate hearing protection has been worn for the whole period since the start
of the shift. As long as this test result is no worse than the last one it can
be accepted, otherwise a longer noise-free period will have to be arranged
for a retest. These compromise arrangements must not be used for the first
baseline audiogram, which will be used for all future comparisons. The operator
of the audiometer may be an industrial nurse or a technician, and they may
conveniently manage the system of recall for regular testing as well. Special
training courses in audiometric techniques are available for these staff, so
that their competence in this area cannot be questioned at some future time.

When a new programme is being set up it is prudent to test the whole
workforce exposed to noise. Evidence from the noise survey can be used subsequently to identify the most suitable intervals for regular retesting of workers
from the differing noise exposure level areas. For instance, it may be felt
that those staff whose DAI L_{Aeq} would exceed 85 dB(A) were it not for hearing
protection, should be tested annually for the first three years and two-yearly
thereafter. This assumes that no untoward hearing changes have occurred.

Figure 6.6 *Audiometric booth for use in ambient noise levels of greater than 45 dB(A)*
(Reproduced by permission of I.A.C. Ltd.)

Those workers whose DAI L_{Aeq} would exceed 95 dB(A) were it not for hearing protection, could be tested annually on a permanent basis, but six-monthly if any hearing deterioration occurred. Any failure to attend for audiometry should be followed up, so that no workers are lost from the scheme.

Interpretation of the audiogram

The results of audiometry must be interpreted in relation to the subject's age, since some high tone loss is usual with normal ageing effects. An example of an age-weighted categorization table is given in Table 6.7. This table and method comes from the UK Health and Safety Executive discussion document *Audiometry in Industry*.[13] The method depends on the notion of summing the audiometric threshold levels for low frequencies 500 Hz, 1000 Hz and 2000 Hz, and high frequencies, 3000 Hz, 4000 Hz and 6000 Hz, for each ear separately. The audiogram is now effectively summarized by a low and high frequency value for each ear. These values are then compared with the table for the age of the test subject. The method allows for five categories of result.

Category 1 includes audiograms in which the sums of the hearing levels, either for low or high frequencies, show an increase of 30 dB or more when compared with the immediately preceding examination, or 40 dB when the time since the previous audiogram exceeds three years. This category could be paraphrased as 'worse than last time'.

Category 2 applies when differences in the sums of the hearing levels between the two ears exceed 45 dB for the low frequencies or 60 dB for the high frequencies. This category can be paraphrased as 'one ear worse than the other'.

Category 3 is based on the 'referral' levels given in the table. Results which

Table 6.7

Chart for categorization of hearing levels

| Age in years | Sum of hearing levels | | | |
| | 500 Hz 1000 Hz 2000 Hz | | 3000 Hz 4000 Hz 6000 Hz | |
	Warning level	Referral level	Warning level	Referral level
20–24	45	60	45	75
25–29	45	66	45	87
30–34	45	72	45	99
35–39	48	78	57	111
40–44	51	84	60	123
45–49	54	90	66	135
50–54	57	90	75	144
55–59	60	90	87	144
60–64	65	90	100	144
65 and over	70	90	115	144

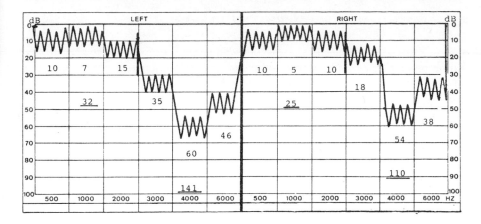

Figure 6.7 *Audiogram – subject's age 43: result; category 3: referral (Lt. high freq)*

are sufficiently below normal for age to come into this category, justify referral of the worker to the designated medical practitioner for examination. This category could be paraphrased as 'bad enough to see a doctor'.

Category 4 is based on the 'warning' levels given in the table. Results which fall in this 'slightly below normal for age' category justify the worker being warned of his or her borderline hearing status. Attention should be given to whether or not hearing protection is being used properly, and to the possibility of significant noise exposure out of work time, that is, hobby noise exposure. This category could be paraphrased as 'just below normal, see again soon'.

Category 5 is to cover those workers who do not fall into any of the previous categories. This category can be paraphrased as 'normal hearing for age'.

An example of a self-recorded audiogram with its resulting categorization is shown in Figure 6.7. The threshold for each frequency is taken as the midpoint of the excursion of the tracing produced for that frequency.

The precise action to be taken for each of the groups now separated into categories should be decided by the medical advisors together with the health and safety team. Generally it is felt that categories 1, 2 and 3 should be referred for a medical opinion. Any monitoring programme of this kind will show up a number of hearing losses which are not caused by noise. Some of these will be well known to the workers involved, others may be discovered for the first time but may still not be the result of noise at work. The hearing loss may be curable, or may be a side effect of some other medical condition that needs treatment. This categorization scheme is not foolproof in practice since it is not sensitive to small changes in consecutive audiograms, but it is valuable while clearer guidance from new legislation is still pending.

Employees' access to test results

Some people may request copies of the results for independent interpretation, or for a new employer. A policy must be agreed in advance on access to the audiogram results.

A common experience once monitoring audiometry has been adopted is that workers become more interested in their own hearing. They will want to know how their test result compares to normal, and usually the test process serves to undermine their possible earlier complacence that since they were used to noise it would not harm them. The value of the health and safety educational spin-off from audiometry is considerable, adding to its attraction as a means of monitoring the effectiveness of the hearing conservation programme.

References

1   Department of Employment (1972) *Code of Practice for the Reduction of the Exposure of Employed Persons to Noise*, HMSO, London.
2   Health and Safety Commission (1987) *Prevention of Damage to Hearing From Noise at Work*, Draft proposals for regulation and guidance, HMSO, London.
3   BS 5108 (1974) *Method of Measurement of Attentuation of Hearing Protectors at Threshold*, British Standards Institute, London.
4   Martin, A.M. (1977) The Acoustic Attenuation Characteristics of 26 Hearing Protectors Evaluated Following the British Standard Procedure, *Ann. Occup. Hyg.*, Vol. 20.
5   Abey-Wickrama, M.F., A'Brook, F.E., Gaton, I.G. and Herridge, C.F. (1969) Mental Hospital Admission and Aircraft Noise, *Lancet*, December.
6   Broadbent, D.E. (1957) Effects of Noise on Behavior, chap. 10 in Harris, C.M. (Ed.) *Handbook of Noise Control*, McGraw-Hill, New York.
7   Broadbent, D.E. (1962) *Effects of Noise on Performance and Productivity*, Paper E–4 National Physical Laboratory Symposium No. 12, The control of noise, HMSO, London.
8   Fox, J.G. (1971) Background Music and Industrial Efficiency – a Review, *Applied Ergonomics*, Vol. 2, p. 70.
9   Hockley, G.R.J. (1969) Noise and Efficiency in the Visual Task, *New Scientist*, May.
10   Hockley, G.R.J. (1970) Effect of Loud Noise on Attenuational Selectivity, *Quarterly Journal of Experimental Psychology*, Vol. 22, p. 28.
11   Weston, H.C. and Adams, S. (1935) *The Performance of Weavers under Varying Conditions of Noise*, Report No. 70, Industrial Health Research Board, HMSO, London.

12 Shipton, M.S. and Robinson, D.W. (1975) *Ambient Noise Limits for Industrial Audiometry*, Acoustics Report AC 69, National Physical Laboratory, UK.

13 Health and Safety Executive (1978) *Audiometry in Industry*, discussion document, HMSO, London.

Chapter 7

NOISE AND VIBRATION IN BUILDING SERVICES

Diane M. Fairhall, Institute of Environmental Engineering, South Bank Polytechnic

The problem for the building services design engineer when dealing with noise in buildings is twofold: first, to ensure that the building is not unduly affected by ambient noise around the site and, second, to prevent as little extra noise as possible being contributed from the equipment and plant necessary to service the building. This chapter, which is primarily concerned with noise emanating from building services equipment and available methods of control.

The essence of the engineer's task is to achieve some predetermined criterion for background noise levels. Recommended criteria, for example noise rating (NR) and noise criteria (NC) curves, are published in various design guides such as those of the Chartered Institution of Building Services Engineers[1] and the American Society of Heating, Refrigeration and Airconditioning Engineers.[2] Although it is the engineer's responsibility to ensure that criteria are met, this task is frequently made difficult due to the lack of information about sound power outputs of machinery and inadequate prediction techniques. It is the object of this chapter to assist the engineer with noise control problem-solving and in establishing the best practicable means to achieving noise criteria.

Noise in air distribution systems

Perhaps the most commonly experienced source of noise from building services is the air-conditioning or mechanical ventilation system. Frequently the provision of forced ventilation or air conditioning is due to the necessity of using sealed double glazing to prevent the ingress into the building of modern high levels of external traffic noise. Although the fan as prime mover is a main source of noise, other potential sources should not be overlooked. Aerodynamic noise caused by air flow over obstructions in the duct such as volume control dampers is often the reason for noisy systems, and poor duct system design (frequently forced upon the engineer due to lack of space) may further exacerbate the problem. It is good practice to outline potential noise problems to the design team at an early stage in the project so that they may be avoided or eliminated.

Fan noise

All fans generate noise by means of a variety of mechanisms and on occasion will give rise to unacceptable levels of vibration. The passage of the blades past a fixed point (such as the outlet duct) will produce noise tonal in character at the blade passage frequency and its harmonics. There will also be noise due to the forces exerted by the fan blades on the air, and the finite thickness of the impeller blades.

Noise will be generated by turbulence due to shear forces in the air, and vortex-shedding noise will also be apparent due to flow separation at solid boundaries. Turbulent wakes impinging on obstructions in the airstream will cause aerodynamic noise, and vibration-related noise originating from the fan drive and motor may radiate from the fan casing and also propagate along the inlet and outlet ducts.

Any obstruction close to the impeller may greatly increase aerodynamic noise. The worst effects are produced whenever a disturbance is periodic, such as obstructions upstream of an axial fan impeller producing a turbulent wake which is struck by the impeller blades. This will be amplified if the number of stator blades is equal to the number of rotor blades. Problems also arise where obstructions downstream of a centrifugal fan impeller (such as the fan cut-off) are hit by successive velocity peaks from the impeller.

The mechanisms described above generate noise of differing character: turbulence will produce broad-band noise (noise at most frequencies in the audible range) and blade rotation and vibrations will produce pure tones (noise at discrete frequencies and their harmonics). If the tonal noise level is below the broad-band noise level, it may usually be ignored, but if it is 3 dB higher than the mean level of the adjacent octave bands, it will be readily perceptible to the human ear and will require treatment.

For a well-designed fan operating at peak aerodynamic efficiency, noise due to vortex shedding is the most prominent source of sound, but at poor efficiency blade rotation, interference and turbulence noise will also become significant. Work by Curle[3] has shown that noise due to flow separation (dipole source) may be expressed:

$$W \propto \rho u^6 c^{-3} d^2 f(\text{Re}) \qquad \text{watt}$$

where ρ is the air density, kg/m³
       u is the flow velocity, m/s
       c is the velocity of sound, m/s
       d is the significant dimension, m
       $f(\text{Re})$ is a function of the (dimensionless) Reynolds number
       u/c is the (dimensionless) Mach number.

For a fan, u is the impeller tip velocity, and d is the fan diameter. Hence it will be seen that the expression simplifies to:

$$W \propto \rho M^3 N^3 D^5 f(\text{Re}) \qquad \text{watt}$$

where $N^3 D^5$ is proportional to the impeller power of the fan. Fan sound power output is therefore proportional to the product of impeller power and varying functions of the Mach and Reynolds numbers. Empirically it is found that the index of u is nearer 5 to 5.5 for fans than the theoretical value of 6, which is considered to be a Reynolds number effect.

The noise produced will radiate into the airstream and travel not only in the direction of the airflow but back towards the fan inlet in approximately equal quantities, although many fans have directional characteristics resulting in outlet levels several decibels higher than those at inlet.

Since a fan will generate the minimum of noise when operating at peak aerodynamic efficiency, many problems may be avoided by correct fan and system matching. Fan sound power levels quoted by manufacturers and calculated from formulae assume that the fan is operating at this maximum efficiency, and levels may be increased to 5 to 10 dB if this is not the case.

It is sometimes necessary to install systems with variable duty, and for a single-fan system where the range is small the maximum and minimum duties should be either side of the peak efficiency point. This may be achieved by the use of vaneaxial fans since the efficiency of this type of fan has a fairly broad peak in the region corresponding to minimum noise generation.

Fans may generate noise levels up to 20 dB louder when stalled (due to severe flow separation) so if the range of duty is large consideration should be given to variable pitch and speed or the provision of two fans instead of one.

If there is a choice between fans with similar efficiencies and duties, the one with the lower blade tip speed will be quieter. Physical size will depend on the space available for installation and maintenance but it is important to limit the discharge velocity, preferably to less than 10m/s.

The general shapes of the sound power spectra of centrifugal and axial flow fans are shown in Figure 7.1. Although centrifugal fans appear quieter than axial flow models, they produce more low frequency noise. Since low frequencies are usually more difficult to attenuate than higher ones, and the cost of absorptive treatment may be high, it is sometimes more cost effective to install a higher-speed, smaller diameter axial flow fan fitted with an attenuator rather than a slower-speed centrifugal.

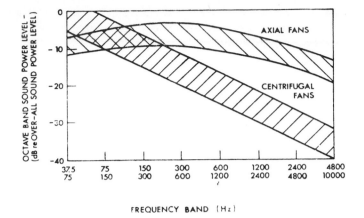

Figure 7.1   *Typical fan noise spectra*
(Reproduced from Section B12 of the CIBSE Guide, by permission of the
Chartered Institution of Building Services Engineers.)

Fan installation also plays an important part in potential noise generation
problems and will be discussed in a later section of this chapter.

Fan sound power output

Manufacturers' data   When estimating the likely sound power distribution
in ducted air systems, it is necessary to know the octave band sound power
level spectrum of the fan. BS 848 Part 2, *Fan Noise Testing*[4] details measurement procedures to be adopted for the determination of fan sound power
output and is adhered to by reputable manufacturers. Such manufacturers
publish noise data in their advertising literature as sound power spectra, and
these may be used in calculations. Other manufacturers either give no noise
data at all or provide them in a form which is of little use to the engineer.
Octave band sound pressure level spectra are of use only where the conditions
of test are also known, that is, the type of environment used for the test,
fan mounting conditions and the distance from the fan to the point of measurement. Armed with this information and a copy of BS 848 Part 2, the engineer
may then work back to the octave band sound power spectrum and use this
in calculations.

Estimation procedures   It is obviously much better to acquire power level
spectra directly from the manufacturer, but occasions arise where this is not
possible, for instance at the preliminary design stage when it may not be
known which manufacturer's fan will be used. Several empirical formulae
exist for estimating fan noise based on the aerodynamic performance, incorporating fan static pressure, delivery volume and motor power. The most useful

is probably that relating sound power to fan static pressure and delivery volume, since these parameters will be known early in the design process, whereas fan motor power will not.

$$L_w = 10\lg Q + 20\lg p + k \qquad \text{dB re 1pW} \tag{7.1}$$

where L_w is the octave band sound power level, dB re 1 pW
Q is the delivery volume, m³/s
p is the fan static pressure, Pa
k is a constant which varies with frequency.

Values of k for different types of fan are given in Table 7.1

As mentioned earlier, fans also generate tonal noise at the blade passage frequency and its harmonics. To account for this a tonal correction should be added to the octave band in which the blade passage frequency occurs. The magnitude of the correction depends on the type of fan and typical values may also be obtained from Table 7.1.

Table 7.1

Fan spectrum correction factors

| Fan type | Octave band centre frequency, Hz | | | | | | | | Tonal correction |
| | 63 | 125 | 250 | 500 | 1k | 2k | 4k | 8k | |
				'k' value, dB					
Centrifugal									
Backward curved:	30	29	30	25	17	15	8	3	5
Forward curved	35	32	27	23	19	13	10	5	5
Radial:									
low pressure	41	36	34	27	26	12	12	11	6
high pressure	44	37	36	34	33	29	27	26	7
Axial									
High hub ratio	32	33	34	35	34	32	28	27	4
Low hub ratio	30	30	31	32	32	29	26	23	4
Propeller									
Curved blade	37	35	34	33	32	30	25	17	5
Aerofoil	37	41	46	45	44	41	35	30	5
Heavy duty	46	48	49	51	50	49	47	42	5

$$\text{Blade passage frequency} = \text{number of blades} \times \text{fan speed} \qquad (7.2)$$

where the fan speed is in revolutions per second.

Example   A centrifugal fan with sixteen backward curved aerofoil blades delivers $5\text{m}^3/\text{s}$ against 1 kPa when running at 1000 rpm. Estimate the sound power spectrum of the fan.

Using equation 7.1:

$$L_w = 10\lg(5) + 20\lg 1000 + k \qquad \text{dB re 1pW}$$
$$= 67 + k \text{ dB re 1pw for each octave band}$$

Using equation 7.2:

$$\text{Blade passage frequency} = 16 \times 1000/60 = 267 \text{ Hz}$$

This frequency falls in the 250 Hz octave band, and the tonal correction should be added there. The k value and tonal correction are obtained from Table 7.1, and so the approximate sound power spectrum will be:

	Octave band centre frequency, Hz							
	63	125	250	500	1k	2k	4k	8k
Basic level	67	67	67	67	67	67	67	67
k value	26	24	23	19	17	15	8	3
Tonal correction			5					
Power level dB re 1pw	93	91	95	86	84	82	75	70

These calculated values are minimum levels corresponding to fan operation at maximum efficiency and should be treated as such by the engineer. Should the fan not be operating at optimum efficiency, the corrections given in Table 7.2 should be added.

Table 7.2
Correction for fan efficiency

Efficiency, % of peak	Correction, dB
90–100	0
85–89	3
75–84	6
65–74	9
55–64	12
45–54	15

Fan installation

The quality of fan installation is of great importance since poor installation may increase the loudness by a factor of two or more. To avoid this the airflow at inlet to the fan should be as uniform as possible. Poor inlet conditions not only generate excessive noise but also contribute to fan inefficiency and system pressure losses. Intelligent system design will therefore not only reduce noise at source but also improve system efficiency and reduce running costs.

Because the fan will radiate sound from both casing and inlet, air-conditioning and ventilation plant should not be sited near noise-sensitive areas. Fans will frequently be situated in plant rooms, and so noise break-out from the plant room should be considered at an early stage. Attenuators may be used on the fan inlet, and the fan casing may be lagged, but these measures may not be sufficient to limit noise to acceptable levels and the fan may need to be enclosed (see Chapter 5).

The flow reaching the fan should be as uniform and laminar as possible, and any fitting that encourages turbulence should be avoided. Bell-mouth or conical inlets should be used on the suction side of the fan and a settling length of at least one duct diameter should be allowed between any duct fitting and the fan inlet. Ductwork should be properly aligned with the fan and flexible connectors pulled taut; they should not be used to cover up poor alignment.

Bends before the fan inlet should have turning vanes fitted and be followed by a settling length of at least one duct diameter. Where positioned after the fan discharge on centrifugal fans, bends with turning vanes should be aligned in accordance with the handing of the fan, taking advantage of the direction of swirl.

Transformation pieces between the fan and the duct should be smooth and gradual with a slope of transmission of one in seven for velocities above 10m/s and one in four below this figure. Area expansions should be limited to about 7 per cent and contractions to about 12 per cent of the original area.

All fans, with attendant motors, should be mounted on resilient mounts as discussed later in this chapter.

Duct break-out and break-in noise

Noise travelling along the duct causes the walls to vibrate, producing a drumming effect. Essentially, the duct walls are acting as loudspeakers and radiate noise to the outside area. In extreme cases break-out may affect the room noise rating. Break-out is less likely to occur where the duct walls are relatively stiff, as in circular ductwork or rectangular ductwork with stiffening straps.

In addition, ventilation ducts are apt to pick up noise when passing through noisy surroundings such as plant rooms. This is known as *break-in*.

Break-out noise   Likely levels of break-out noise are frequently estimated by means of the Allen formula, which considers the worst case.

$$L_{W\text{break-out}} = L_{W\text{duct}} - R + 10\lg(S_D/S) \qquad \text{dB re 1pW} \qquad (7.3)$$

where L_W is the octave band sound power level, dB re 1 pW
S_D is the total surface area of the duct which could radiate sound into the space, m^2
S is the cross-sectional area of the duct, m^2
R is the duct wall sound reduction index, dB.

It should be noted that where a duct runs along a wall or ceiling (as is typical in practice) the directivity index will increase to 2, resulting in a 3 dB increase in level. A duct positioned at the wall and ceiling junction having a directivity index of 4 will result in an increase in the original sound level of 6 dB.

This equation has been used relatively successfully in recent years but in practice is found to overestimate the amount of break-out at low frequency, suggesting that more sound power leaves the duct than enters. It is therefore subject to the limitation that the break-out power is never greater than half the average acoustic power in the duct, that is, $L_{W\text{break-out}} <= L_{W\text{duct}} - 3\,\text{dB}$. Another disadvantage of this formula when applied in practice is that it provides no means of calculating the sound pressure level in the room due to the break-out power.

An equation for calculating the room sound pressure level has been produced by Ver,[5] and yields results that agree well with levels measured on site. This includes a correction term for the attenuation of sound power inside the duct over the length exposed in the room.

$$L_{p(r)} = L_{W\text{duct}} + 10\lg(1/2\pi r + 4L/A) + 10\lg(P/S) + 10\lg(n) - R + C \qquad (7.4)$$

where $L_{p(r)}$ is the sound pressure level in the room at distance r measured perpendicular to the duct axis, dB re 20 μPa
$L_{W\text{duct}}$ is the sound power level of the sound entering the duct at the room boundary, dB re 1 pW
r is the distance from the duct, m
L is the length of duct in the room, m
A is the total absorption in the room, m^2
P is the perimeter of the duct, m
S is the cross-sectional area of the duct, m^2
n is the directivity factor,
   = 1 if the duct is not close to any reflecting surface
   2 if the duct is close to one reflecting surface
   4 if the duct is close to the intersection of two reflecting surfaces
C is the correction term for attenuation inside the duct, given by:

$$C = 10\lg[(1 - 10^{-(\delta L_W)/10})/(0.23 \times \delta L_W)]$$

where δL_W is the change in the sound power level over the duct length.

The sound reduction index will vary with frequency and the break-out level should be calculated for each octave band. Values of sound reduction index for some typical duct constructions are given in Table 7.3.

Crosstalk noise

Ducts often serve as transmission paths between two or more otherwise isolated rooms. This may happen where two rooms are served by adjacent branch take-offs or where several spaces are served from plenum chambers or ceiling voids. Crosstalk may be prevented by installing barriers or baffles between the individual room take-off points or providing crosstalk attenuators before the room outlets.

Frequently, crosstalk is actually flanking transmission through air gaps around the ductwork rather than noise travelling along the ductwork itself. One example of this is where a duct system runs above a suspended ceiling and the partitions between rooms do not continue up through the ceiling void to the underside of the slab. In this case the sound is unlikely to be transmitted along the ductwork, but simply enters the ceiling void and is transmitted to adjacent rooms. Flanking crosstalk may be eliminated by sealing any air gaps around ducts with mastic or ensuring that partitions are continued up to the underside of the slab and all edges sealed to the surrounding structure.

Table 7.3

Sound reduction indices for common duct materials

Construction	Octave band centre frequency, Hz							
	63	125	250	500	1k	2k	3k	4k
1.5mm lead sheet	22	28	32	33	32	32	33	36
3.0mm lead sheet	25	30	31	27	38	44	33	38
20g stiffened aluminium	8	11	10	10	18	23	25	30
22g galvanized steel	3	8	14	20	23	26	27	35
20g galvanized steel	3	8	14	20	26	32	38	40
18g galvanized steel	8	3	20	24	29	33	39	44
16g galvanized steel	9	14	21	27	32	37	43	42
18g fluted steel	25	30	20	22	30	28	31	31

Attenuation in duct systems

In any duct system, whether or not it incorporates attenuation devices, only a proportion of the acoustic energy generated by the fan and fittings will eventually be radiated at terminations. As sound propagates along the duct from the fan towards the room outlets, losses of acoustic energy will occur due to various physical processes. At the same time, acoustic energy may be generated within systems generally due to flow separation and vortex shedding. However, at certain duct elements where losses are in excess of locally generated noise, there will be a net attenuation. The engineer must take account of this fact in design calculations to prevent over-design and the associated costs.

Attenuation in straight ducts   If the duct walls were perfectly rigid there would be little energy loss, but in practice the fluctuating acoustic pressure sets the duct walls in motion converting acoustic energy into mechanical energy. Some of this is dissipated as heat in the walls themselves, and some is re-radiated as sound to the surrounding environment. Whilst this is likely to be a problem in the affected space, it results in a loss of acoustic energy within the duct itself.

At low frequency, more attenuation per metre is obtained in rectangular section ducts than circular ones since the walls tend to be less rigid than those of circular ducts of similar size. In the low frequency range unlined circular ducts provide negligible attenuation per metre, only long duct runs being significant in system design. At high frequency the attenuation per metre for both configurations is similar, since it depends on the mass law, that is, the amount of sound breaking out is governed by the sound reduction index of the duct material, which will be similar for both types of duct. It has been shown that flat oval ductwork gives values of attenuation lying between the rectangular and circular values, but only the ASHRAE guide publishes data. Typical values of attenuation in straight ducts are given in Table 7.4.

Attenuation at elbows and bends   Mitred bends without turning vanes reflect some of the sound back towards the source and this can make significant contributions to attenuation at certain frequencies. The greatest effects occur in 90° bends at the frequency whose wavelength is twice the duct width. Radiused bends provide little or no attenuation (see Table 7.5).

Attenuation at divisions of flow   It has been generally assumed that the acoustic energy divides in direct proportion to the areas of the branch ducts, so that the amount of acoustic energy is less in any branch than in the source duct. It is now felt that this assumption is fundamentally incorrect, since changes in the acoustic impedance at a branch may cause some reflection of acoustic energy. Brown and Stewart[6] have shown that the configuration of the junction

Table 7.4

Attenuation in straight sheet metal ducts

(Reproduced from Section B12 of the CIBSE Guide, by permission of the Chartered Institution of Building Services Engineers.)

Section	Mean duct dimension or diameter (mm)	Attenuation (dB/m) Octave band centre frequency (Hz)		
2		125	250	500 and above
Rectangular	Up to 300	0.6	0.5	0.3
	300–450	0.6	0.3	0.3
	450–900	0.3	0.3	0.2
	Over 900	0.3	0.2	0.1
Circular	Up to 900	0.1	0.1	0.1
	Over 900	0.03	0.03	0.06

Table 7.5

Attenuation provided by 90° mitred bends without turning vanes

(Reproduced from Section B12 of the CIBSE Guide, by permission of the Chartered Institution of Building Services Engineers.)

Minimum duct side (mm)	Attenuation (dB) Octave band centre frequency (Hz)						
	125	250	500	1000	2000	4000	8000
150	0	0	1	6	7	4	4
300	0	1	6	7	4	4	4
450	0	3	7	5	4	4	4
600	1	6	7	4	4	4	4
750	3	7	5	4	4	4	4

will have an effect on the amount of attenuation in each branch, that is, a Y-shaped or swept junction is adequately described by current theory, but a T-shaped junction provides extra attenuation comparable with that of a

90° bend of similar size. Research into this phenomenon is currently in progress, but until results are available the present method may be used for estimation purposes. The attenuation will be:

$$\text{Attenuation} = 10 \lg (A_1 + A_2)/A_1 \quad \text{dB} \qquad (7.5)$$

where A_1 is the area of the branch of interest, m²
       A_2 is the area of the other branch, m².

Should there be a third branch at the junction of area A_3, this would be added to the sum of A_1 and A_2. Note that the area of the main duct before branching does not appear in the equation.

End reflection losses When a duct opens abruptly into a large space such as a room or perhaps a plenum chamber, some of the sound energy will be reflected back toward the source. This is due to the change in the acoustic impedance seen by the sound propagating along the duct. The longer the wavelength (low frequency) the greater the proportion of reflected sound, hence the greater the end reflection loss. At high frequency an impedance change is seen only at small openings because sound at high frequencies becomes beamed along the axis of the duct and approximates a parallel beam so that large openings do not affect the transmission of the wave.

Because low frequencies are the most difficult to attenuate in a duct system, end reflection attenuation is very useful to the engineer and careful duct design may maximize attenuation by this phenomenon. However, overestimation of this loss will result in higher levels than expected in the room the terminal serves, and expensive remedial treatment will be necessary.

It should be noted that the above applies only when the termination is at the end of a duct run; where the opening is on the side different values of end reflection attenuation should be expected. The CIBSE guide suggests that in these cases 50 per cent of the normal value should be allowed for. However, recent work by the author[7,8] has shown that for side openings the amount of end reflection attenuation cannot be predicted by current theory, and published data are not applicable here. Until further information is available it is recommended that no allowance be made for end reflection attenuation where the opening is on the side of a duct.

Empirical end reflection losses for use in the case of terminations on the end of a duct are given in Figure 7.2

Example Consider the acoustic index run of a ventilation system shown in Figure 7.3. Calculating for the 1 kHz octave band only, the attenuation due to the combination of elements will be as shown in Table 7.6.

This is the total attenuation in the 1 kHz octave band, and should be substracted from the fan sound power in this octave band to give the estimated sound power radiated from the diffuser.

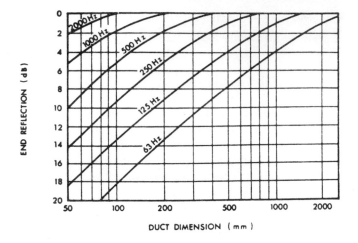

Figure 7.2   *Theoretical value of end reflection attenuation*
(Reproduced from Section B12 of the CIBSE Guide, by permission of the
Chartered Institution of Building Services Engineers.)

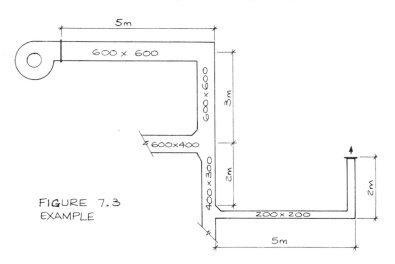

Figure 7.3   *Acoustic index run of a ventilation system.*

Velocity-generated noise

Any obstruction in a duct will generate noise when air flows over or around
it. The greater the flow speed, the greater the sound power generated, and,
for a doubling of velocity. Curle's equation predicts an increase in sound
level of up to 18 dB. Engineers usually expect, or are familiar with, noise

Table 7.6

Attenuation calculation for system in Figure 7.3

Duct element	Calculated from	Attenuation dB (1 kHz)
5m of 600 × 600mm duct	Table 7.4	1.0
Bend 600 × 600mm	Table 7.5	4.0
3m of 600 × 600mm duct	Table 7.4	0.6
Branch	Equation 7.5	4.8
2m of 400 × 300mm duct	Table 7.4	0.6
Branch	Equation 7.5	6.0
5m of 200 × 200mm duct	Table 7.4	1.5
Bend 200 × 200mm	Table 7.5	6.0
2m of 200 × 200mm duct	Table 7.4	0.6
End reflection	Figure 7.2	0.0
Total attenuation		25.1

from volume control dampers and diffusers, but should not forget such things as fire damper curtains in the airstream, or the duct joints themselves. To avoid problems, ductwork should be manufactured to DW/142[9] and fire dampers that do not have part of their curtain in the airstream should be selected.

Velocities in duct systems should be kept as low as possible to prevent vortex-shedding noise from duct elements. If velocities are low, the velocity-generated noise will be less than the level of fan noise in the duct and may be neglected. 'Rule of thumb' guidelines for 'quiet' velocities are:

1   10m/s in main risers.
2   7m/s in branch ducts and risers.
3   4m/s in ducts serving room outlets.
4   2m/s face velocity at terminals.

If these guidelines are followed there should be no significant noise regenerated by duct fittings (sometimes called self-noise). Above these velocities the effect of aerodynamic noise must be allowed for. Both the CIBSE Guide[1] and the ASHRAE Handbook[2] contain procedures for the estimation of velocity-generated noise. Work by Holmes[10] for the Heating and Ventilation Research Association (now Building Services Research and Information Association) also produced algorithms for estimating noise generated by various duct fittings. The approach adopted here will be that of CIBSE, being most typical

Table 7.7

Corrections for low turbulence duct fittings

Duct fitting	Value of c (dB)	Remarks	Octave band power level corrections (dB)						
			Octave band centre frequencies (Hz)						
			125	250	500	1000	2000	4000	8000
Straight duct	−10	No internal projections	−2	−7	−8	−10	−12	−15	−19
90° radiused bend	0	Aspect ratio ≤ 2:1 Throat radius ⩾ $w/2$	−2	−7	−8	−10	−12	−15	−19
90° square bend with turning vanes	+10	Close spaced short-radius single-skin turning vanes	−2	−7	−8	−10	−12	−15	−19
Gradual contraction	+1	Area ratio 3:1 A and v taken for smaller duct	0	−10	−16	−20	−22	−25	−30
Sudden contraction	+4	Area ratio 3:1 A and v taken for smaller duct	0	−10	−16	−20	−22	−25	−30
Bufferly damper	−5	A and v apply to minimum free area at damper	−3	−9	−9	−10	−17	−20	−24

(Reproduced from Section B12 of the CIBSE Guide, by permission of the Chartered Institution of Building Services Engineers.)

of British practice and also the method adopted by most commercial computer software producers in the UK.

The overall sound power level generated at an element is approximately:

$$L_w = 10\lg A + 60\lg u + c \qquad \text{dB re 1pW} \tag{7.6}$$

where A is the minimum flow area of the fitting, m^2
u is the maximum velocity through the fitting, m/s
c is a constant that depends on the individual fitting and flow turbulence, and may be found in Table 7.7.

Octave band correction factors may also be found in this table.

For grilles and diffusers, the following equations may be used. Octave band correction factors are found in Table 7.8.

Grilles:
$$L_w = 10\lg A + 60\lg u + 17 \qquad \text{dB re 1pW} \tag{7.7}$$

Diffusers:
$$L_w = 10\lg A + 60\lg u + 32 \qquad \text{dB re 1pW} \tag{7.8}$$

It should be remembered when using these equations that the velocity to be used in the calculation is not always the duct flow velocity but the local velocity at the fitting, which, at a constriction, will be higher than the duct velocity.

Table 7.8
Corrections to overall sound power level for diffusers and grilles
(Reproduced from Section B12 of the CIBSE Guide, by permission of the Chartered Institution of Building Services Engineers.)

$\dfrac{\sqrt{A}}{u}$ (s)	Corrections (dB) Octave band centre frequency (Hz)						
	125	250	500	1000	2000	4000	8000
0.01	−10	−8	−7	−6	−6	−7	−12
0.02	−8	−7	−6	−6	−7	−12	−21
0.04	−7	−6	−6	−7	−12	−21	−30
0.06	−6	−5	−6	−10	−17	−26	−35
0.08	−6	−6	−7	−12	−21	−30	−40
0.10	−5	−6	−8	−15	−24	−33	−43

Example   A 300×300mm square grille having a free area of 50 per cent delivers $0.5\text{m}^3/\text{s}$ of air. Estimate the sound power spectrum of the noise regenerated at the grille.

$$\text{Free area of grille} = 0.3 \times 0.3/2 = 0.045\text{m}^2$$

$$\text{Velocity through grille} = 0.5/0.045 = 11.11\text{m/s}$$

Using equation 7.7:

$$L_w = 10\lg(0.045) + 60\lg(11.11) + 17 \quad \text{dB}$$
$$= 49.3 + 17 = 66.3 \text{ dB re } 1\text{pW}$$

	Octave band centre frequency, Hz						
	125	250	500	1k	2k	4k	8k
Basic level (equation 7.7)	66.3	66.3	66.3	66.3	66.3	66.3	66.3
Correction (Table 7.8)	−8.0	−7.0	−6.0	−6.0	−7.0	−12.0	−21.0
Total	58.3	59.3	60.3	60.3	59.3	54.3	45.3

This is an extreme example illustrating the effect of high velocities through grilles and is not intended to represent realistic design practice.

Estimation of room sound levels

The foregoing estimation procedures will enable the engineer to calculate the likely sound power level at any point in the duct system. Before the design of the system can be finalized, the engineer must determine whether the room criteria will be achieved with the present design or if additional noise control methods must be used. To make this decision it is necessary to calculate the room sound pressure levels due to any air distribution system noise entering the served spaces. The sound pressure level in any room will depend not only upon the sound entering from the duct system, but also upon the absorption properties of the room itself.

Sound fields in rooms

The direct field   The sound field in a room due to any source will be a combination of direct and reverberant components. The direct component is that arriving at the listener from the source before any reflection has occurred, and depends on the position of the source relative to the room boundaries, and the distance of the listener from the source. The sound pressure level due to the direct field component is given by:

$$L_{p(d)} = L_W + 10\lg Q/4\pi r^2 \quad \text{dB re } 20\ \mu\text{Pa} \tag{7.9}$$

where Q is the directivity factor of the source
   r is the distance from the source, m
   L_W is the sound power radiated from the source, dB re 1pW.

It may be seen from equation 7.9 that the direct sound pressure level is inversely proportional to the square of the distance from the source, and this represents a decrease of 6 dB every time the distance from the source is doubled.

The directivity factor depends on the position of the source in the room. A source of sound power W radiating into a sphere will, over its surface at distance r, produce a sound intensity I. If the source is restricted to hemispherical radiation, for instance placed on the floor, the intensity at distance r will be doubled since the surface area is halved, and the factor of 2 by which it has increased is called the *directivity factor*, Q. A source positioned at the junction of two planes, such as a wall and ceiling, has a directivity factor of 4, and if placed in a corner it has a directivity factor of 8.

Reverberant field   The reverberant field is composed of the sound remaining after reflection has taken place at the room boundary, and the reverberant level will depend on the amount of absorption present in the room. It reaches a steady state level when the amount of sound energy entering the room is equal to the amount absorbed at the room surfaces. The sound pressure level due to the reverberant component is given by:

$$L_{\text{pr}} = L_W + 10\lg[4(1 - a)/S\alpha] \quad \text{dB re } 20\ \mu\text{Pa} \tag{7.10}$$

where S is the total area of all room surfaces, m²
   α is the mean absorption coefficient of the room surfaces.

Noise control in the duct

System design and installation

The simplest method of achieving low duct noise levels is to design the system to avoid poor layout and excessive velocities. This will not always be possible due to lack of space and cost of duct materials but it is a valuable exercise to compare the costs and noise generation properties of larger ducts with those of smaller ducts with silencers.

Systems should be kept simple, using a minimum of fittings such as branches and bends, and joints between fittings should be smooth with no internal projections. Branch take-offs should be made gradual and smooth as should changes in dimensions or cross-sectional area and transitions between fittings. Bell-mouth or flared fittings on intake and exhaust reduce local flow velocities and turbulence and hence the noise levels.

Volume control dampers should be avoided where possible by designing the system to be self-balancing, but where they are necessary they should be positioned at a distance well upstream of any room outlet. This distance should be at least five duct diameters, although ten is better, and lining the last five duct diameters before an outlet with 25mm-thick acoustic absorbent material will give added protection against noise entering the room.

Dampers positioned directly behind grilles and diffusers should be used for fine trimming only, a maximum pressure drop of 20 to 30 Pa. Noise level increases with pressure drop and an increase in level of up to 10 dB may occur with each doubling of pressure drop. If low noise ratings are to be achieved, that is, NR 30 or less, the grille dampers must be fully open, and flow control achieved further back from the grille. If large pressure drops are required, then two (or more) dampers, separated by at least five duct diameters should be used to reduce the pressure in stages.

Devices that are likely to disturb the flow such as bends, branches and changes in section should be well separated since interactions between fittings, although at present not fully investigated, are accepted to increase levels to above the combination of levels that each element would generate alone. Using the five- to ten-diameter rule will enable the flow to settle before meeting the next element.

Ductwork of large dimensions may be stiffened to prevent alternate 'ballooning' and collapsing of the duct and the associated drumming noise. Braces made of metal such as angle iron or folded sheet metal lined with resilient material such as neoprene may be attached to the perimeter of the duct to give extra support.

Duct supports and hangers should be lined with resilient material such as felt or neoprene in order to isolate them from the duct. In noise-sensitive areas resilient mounts such as proprietary spring hangers should be used to prevent noise transmission through the supporting structure. Resilient mounting will also allow any expansion or contraction to take place without creaking.

Acoustic treatment

Acoustic treatment of ductwork may be of the resistive type such as packaged silencers and duct linings where sound is attenuated by viscous losses in the absorptive lining, or the reactive type such as the expansion plenum where attenuation relies on partial reflection of the sound back towards the source. A thorough analysis of the proposed duct system will enable problem areas to be identified, and a knowledge of the likely duct levels will facilitate the selection of the appropriate acoustic treatment.

The first stage is to ascertain the critical paths through the system, and to identify the areas where noise from the duct system (including duct outlets and break-out) exceeds the design criterion, and hence to determine the actual attenuation required over the frequency range. This should establish whether the whole system must be treated, local treatment is sufficient or a combination of the two is necessary.

Packaged silencers   In systems where there is no main noise source except for the fan, acoustic treatment may be as simple as installing a silencer at the fan outlet. Packaged units may be obtained ready-made, and may be selected from a range to suit duct size, available space and allowable additional system pressure loss as well as the required attenuation. Reputable manufacturers provide this information in their sales literature, and will also assist the engineer with silencer selection if requested. Figure 7.4 is an illustration of silencers fitted to a system.

Generally, the air passage width and silencer length will determine the amount of attenuation, whilst the total width and height as well as inlet and outlet conditions will determine the pressure loss over the unit. For greater attenuation, reduce the passage width and increase the length of the silencer, and for lower flow resistance increase the number of airways and the height of the unit and ensure that there is smooth transition at the entry and exit ducts. This last is of particular importance where the unit is greater in cross-section than the parent duct.

Figure 7.4   *Silencers installed in an air-conditioning system*
(Reproduced by permission of Industrial Acoustics Company Limited.)

To help reduce noise and vibration passing from the fan to the structure by way of the duct and attenuator supports, it is good practice to use spring hangers on attenuators where they are suspended from the slab. Manufacturers of such supports will assist the engineer with selection if necessary. Flexible connections should be used to isolate the fan from the ductwork or attenuator, and flanges should be properly aligned and the flexible connections pulled taut between them.

If the fan sound power is high, the silencer should be installed as close as possible to the fan to prevent any break-out from the duct. However, in noisy plant rooms where there is a danger of noise from other fans and equipment breaking back into the duct after the silencer, a better position is in the plant room wall. An alternative solution is to insulate the outside of all exposed ductwork with a dense material such as Keen's cement, although this could be expensive. The decision will probably rest on allowable noise levels in the plant room.

For axial flow fans with their greater mid and high frequency noise, in-line attenuators should generally be adequate since absorption is fairly good at these frequencies. These fans are also easy to enclose fully with the motor in the airstream, and this is helpful if quiet plant rooms are required. Centrifugal fans, with their highest levels at the lower frequencies, will almost invariably require a splitter-type silencer to achieve the necessary attenuation.

In some systems it may be necessary to install both first and second stage attenuators to meet criteria. This will be so if there is a significant amount of regenerated noise. With this arrangement, the first stage attenuator is installed at the fan, and the second stage attenuators as necessary behind the terminal units. This sometimes means that the first stage attenuator can be smaller and correspondingly less expensive than a single attenuator silencing the whole system. This arrangement is additionally useful in helping to prevent crosstalk noise between rooms.

Duct linings   In some situations it may be possible to achieve the required attenuation by lining certain sections of the duct. Lining can often be an economic solution because as well as attenuating the sound it may also serve as thermal insulation and introduces only a slight increase in pressure drop.

To be effective acoustically, the lining should be at least 25mm thick, and installed on the inside surface of the duct. This is very effective at the higher frequencies but adequate attenuation at the lower end of the spectrum may need additional or alternative treatment such as expansion plenums. Lined bends are particularly efficient as attenuators since they combine the two mechanisms of absorption and reflection, and sufficient attenuation may be obtained in some systems by adopting this approach.

Duct linings are cheaper than splitter-type silencers, but as the duct increases in size the less the attenuation available per metre length of the duct. This is because the ratio of perimeter to area controls the possible attenuation, and the larger the duct, the smaller this ratio will be. The attenuation of a lined duct may be estimated from:

$$\text{Attenuation} = S\alpha^{1\cdot4}/A \quad \text{dB} \quad\quad (7.11)$$

where S is the total area of absorptive material exposed to the sound, m²
α is the acoustic absorption coefficient of the material
A is the cross-sectional area of the airway, m².

Since the absorption coefficient for any material will vary with frequency, so will the amount of attenuation for a specified section of duct.

Expansion plenum chambers   Plenum chambers (illustrated in Figure 7.5) are frequently incorporated into air distribution systems as settling chambers, and lining these with acoustic absorbent provides a simple means of attenuating fan noise. If the dimensions of the plenum are large with respect to the inlet duct, low frequency attenuation is improved due to reflection effects. Further attenuation may be obtained by staggering the inlet and outlet openings and/or using partial baffles inside the chamber. The attenuation of a lined plenum may be estimated from:

$$L_{W2} = L_{W1} - 10\lg[S_0(\cos\theta/2\pi d^2 + 1/R_c)] \quad \text{dB} \quad\quad (7.12)$$

where S_0 is the outlet duct area, m²
d is the distance from centre of inlet to centre of outlet, m
θ is the slant angle between inlet and outlet
R_c is the 'room constant' of the plenum, given by $R_c = S_t\,\alpha/(1-\alpha)$
S_t is the total internal surface area of the plenum chamber including inlet and outlet duct areas, m²
α is the mean absorption coefficient of the internal surfaces, inlet and outlet duct areas having an absorption coefficient of 1.

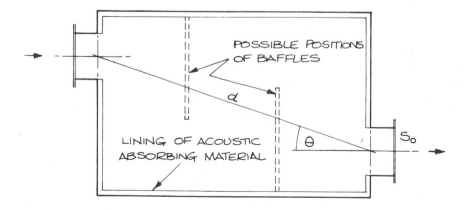

Figure 7.5   *Expansion plenum chamber*

Example   A sheet metal plenum chamber is in the form of a cube of side 1m, and is lined with 25mm of acoustic absorbent having an absorption coefficient of 0.3 in the 500 Hz octave band. If the inlet and outlet ducts, both measuring 200 × 200mm, are positioned centrally in opposite faces of the cube, estimate the attenuation of the chamber in the 500 Hz octave band. Using equation 7.12:

$$S_0 = 0.2 \times 0.2 = 0.04m^2$$
$$d = 1m$$
$$\theta = 0 \text{ (inlet and outlet in line)}$$
$$S_t = 0.95 \times 0.95 \times 6 = 5.415m^2$$
$$\text{Area of absorbent} = 5.415 - 0.08 = 5.335m^2$$
$$a = (5.335 \times 0.3 + 0.08 \times 1)/5.415 = 0.31$$
$$R_c = 5.415 \times 0.31/(1 - 0.31) = 2.43m^2$$

Substituting into the equation:

$$L_{W1} - L_{W2} = -10\lg[0.04 \times (\cos 0/2\pi \times 1 + 1/2.43)]$$
$$= 16 \text{ dB at 500 Hz}$$

It is unlikely that this figure would be achieved in practice, but the equation gives a good estimation of the likely attenuation at each frequency. A practical value of attenuation here would be 2 or 3 dB lower than the predicted value.

Noise in water distribution systems

Water system noise, like air system noise, has several origins. Pumps and attendant motors are analogous to fans, producing primary noise in the system, but additional noise due to boiler parts such as induced or forced draught fans and burners may add to this. The transmission paths are through the water and the pipe walls resulting in airborne noise to the surroundings, and also structureborne through pipe supports to the building, often resulting in re-radiated airborne noise.

As in air systems, valves and other fittings in the pipe system cause obstructions to the flow and produce noise. Control valve noise occurs due to the vibration of components and turbulence in the flow caused by the valve. Pressure fluctuations due to pipe fittings may cause water hammer or allow air dissolved in the water to be driven off and form air pockets leading to cavitation. If velocities are too high then noise may be experienced from the flow itself.

Pipework noise

Noise generated by the fluid flow will depend, as in air systems, on the flow velocity within the pipe. Where flow is turbulent (most practical systems), noise levels will rise with increasing Reynolds number. Valves and fittings generate more noise than straight pipes and exhibit different spectra. However, it is found that partly open valves generate significantly higher noise levels than straight pipes and other fittings, and therefore the overall system noise level will depend mainly on the flow velocity and the rate of flow through valves.

Pump noise

The noise generated by pumps is generated directly proportional to the pressure and the speed and inversely proportional to the size of pump. The noise spectrum is generally broad band in nature, with harmonics due to fluctuating discharge pressure and the fundamental rotational frequency. Spectra for different types of pump will therefore vary widely, some types of pump being inherently quieter than others. Modern, compact in-line pumps are available for a wide range of duties and are relatively silent running.

Hydrodynamic noise due to the turbulent flow is a main component of pump noise and occurs at the blade passage frequency and harmonics as given for fans in equation 7.2, and may be modified by changing the operating conditions and/or the pump geometry. For instance, halving the pump speed for a particular duty may give a noise reduction of up to 12 dB.

Cavitation may substantially increase the total noise level in the system, although at onset appears to attenuate to some degree the blade frequency noise. Cavitation is caused where the local pressure falls near to or below the saturation pressure of the fluid enabling vapour bubbles to form. When the local pressure increases the vapour condenses and the bubbles collapse, causing severe damage to pumps and pipework. Cavitation may also occur when trapped air bubbles similarly collapse, but this occurs more slowly and causes less damage. Cavitation should be avoided due to its damaging effects on the system, not simply to prevent excessive noise. The system should be designed such that the available net positive suction head (NPSH) of the pump is sufficient to prevent cavitation occurring. Further discussion of this topic is beyond the scope of this book, and the reader is referred to one of the many texts on fluid mechanics.

Other instabilities in the fluid flow, and also shaft imbalance, may result in unacceptable amplitudes of vibration. The use of anti-vibration mountings and the isolation of pipework from the supporting structure will frequently be enough to prevent problems arising, but if this is not sufficient it may be necessary to change to a different type of pump.

Valve noise

Control valves in hydrodynamic systems are principal sources of noise, produced partly from mechanical vibration of moving parts and partly from hydrodynamic and aerodynamic phenomena. The main source of mechanical noise is the lateral movement of the valve plug in its seat, sometimes known as valve chatter, and generally of no greater frequency than 1500 Hz. This usually occurs when the valve seat becomes old and worn, and replacement will cure the problem. A secondary source is resonance of a component at its natural frequency, this usually being in the range 3 to 7 kHz. Mechanical noise is unpredictable and may be eliminated by good valve design.

Water hammer may occur due to abrupt changes in flow velocity caused by the valve or sometimes by an abrupt change in pump speed. These sudden changes produce pressure waves in the pipe which are reflected backwards and forwards by the pipe ends, generating 'knocking' noises or hammer and possibly causing damage to the pipework system. Water hammer may be avoided by decreasing the normal flow velocity in the pipe or ensuring that control devices have a smoother response. Flexible pipe sections installed in the system will assist in damping out large pressure fluctuations.

Cavitation, discussed above, may occur at control valves if the flow rate is decreased sufficiently. This is the reason why control devices such as taps and valves are noisier when partly closed than when they are fully open. Valve selection is important in reducing noise from this source, and if the manufacturer's recommended flow rates are adhered to, cavitation noise should not occur.

Machinery vibration and isolation

Most machinery in common use within the building services industry is likely to give rise to some vibration generally due to moving parts. The frequency at which this occurs is called the *frequency of forced vibration,* or *forcing frequency*. At the same time, the particular piece of equipment will have one or more frequencies of natural vibration, that is, the frequency at which it will freely oscillate after being displaced from its equilibrium position. If the forcing frequency coincides with any of the natural frequencies of the machine, resonance occurs and the amplitude of vibration is magnified. If magnification is great enough, physical damage may occur to machine or structure, but in any case the resonant condition will increase the noise output of the system.

For example, with a fan, rotation of the impeller will result in vibration which may be transmitted along ductwork and also to the building structure. The frequency of forced vibration will be equal to the number of revolutions the impeller makes per second. Since most fans are driven by electric motors additional vibrational energy will be experienced from the motor, the forcing frequency here being twice the electrical supply frequency.

To avoid annoyance from vibration and ensuing noise, machinery should be mounted on resilient supports, and not simply bolted directly to the floor slab. Machines with attendant motors should be mounted on a common base, and the base designed for the lowest forcing frequency of the system, hence automatically isolating any higher forcing frequency.

Free vibrations

Undamped systems

Most mechanical systems met in practice obey to some extent the laws of simple harmonic motion. For a system to oscillate, it must be able to store potential energy, that is, it should have some elasticity and it must have some inertia or mass to enable it to possess kinetic energy. Oscillation is the continual interchange of potential and kinetic energy. Simple harmonic motion occurs very frequently and most systems will exhibit at least approximate simple harmonic motion for small oscillations about an equilibrium point.

In practice, an undamped system may be represented by a load mounted on steel springs, which have very little inherent damping. When the load is applied to the springs, the springs will be compressed a certain distance, depending on the load mass, m, and the stiffness, k of the springs. This distance is known as the *static deflection*, δ, of the system. If the load is now pushed down and then released, the system will oscillate about its rest position, and the frequency at which it oscillates is known as the *natural frequency*, f_n, of the system. It can be shown from the theory of simple harmonic motion that the relationship between natural frequency, spring stiffness and mass is given by:

$$f_n = (k/m)^{0.5}/2\pi \qquad \text{Hz} \qquad\qquad (7.13)$$

and similarly the relationship between natural frequency and static deflection can be shown to be:

$$f_n = (g/\delta)^{0.5}/2\pi \qquad \text{Hz} \qquad\qquad (7.14)$$

where g is the acceleration due to gravity, 9.8m/s^2, and if δ is in millimetres, this reduces to:

$$f_n = 15.76(\delta)^{0.5} \qquad \text{Hz} \qquad\qquad (7.15)$$

Example   An uncompressed spring measures 300mm and 250mm when a mass of 50kg is placed upon it. Calculate the natural frequency and the stiffness of the spring.

$$\delta = 300 - 250 = 50\text{mm}$$

$$f_n = 15.76(50)^{-0.5} = 2.229 \text{ Hz}$$

$$2.229 = (k/50)^{5.0}/2\pi$$

$$k = (2.229 \times 2 \times \pi)^2 \times 50 = 9.8073 \text{ kN/m}$$

Damped systems

The motion of most oscillating systems in practice is opposed by dissipative forces such as air viscosity, and energy is removed from the system, usually being converted to heat. The process of removing energy in this way is known as damping, or mechanical resistance, designated c, the units being Ns/m.

There are three possible forms of solution to the differential equation of motion representing different degrees of damping. These are:

1 *Overdamping.* After any disturbance the system reverts to its original position without any oscillation, its response being sluggish.
2 *Critical damping.* After disturbance the system returns to its equilibrium position in the shortest possible time, and without overshoot. The critical damping factor for this condition is designated c_c.

$$c_c = 2 \, (km)^{0.5} \qquad \text{Ns/m} \qquad (7.16)$$

3 *Underdamping.* When a system is underdamped a disturbance results in oscillations having diminishing amplitude with time. This is generally the condition controlling engineering systems. The relationship c/c_c in any system is known as the damping ratio, ζ.

$\zeta > 1$ represents overdamping
$\zeta = 1$ represents critical damping
$\zeta < 1$ represents underdamping.

The damped frequency of natural vibration is given by:

$$f_d = f_n(1 - \zeta^2)^{0.5} \qquad \text{Hz} \qquad (7.17)$$

Generally speaking, for the relatively small values of damping ratio encountered in building services systems (of the order of 0.1) the difference between the damped and undamped natural frequencies is of little practical significance.

Forced vibrations

So far consideration has been given only to systems initially disturbed and left to oscillate freely until they return to their rest position. In practical systems, the force is applied continuously in a sinusoidal manner, sustaining the oscillations at some steady state amplitude. This steady state value is a function of forcing frequency, natural frequency, stiffness and damping. The dynamic magnification factor or amplitude ratio is the ratio of the dynamic amplitude of the system to the amplitude of forced vibration, given by:

$$\frac{x}{(F_0/k)} = \frac{1}{[(1 - (\omega/\omega_n)^2)^2 + (2\zeta\omega/\omega_n)^2]^{0.5}} \qquad (7.18)$$

where x is the dynamic amplitude of vibration, m
F_0 is the maximum value of the applied force, N
ω is the angular forcing frequency, $2\pi f$ radians/s
ω_n is the angular natural frequency, $2\pi f_n$ radians/s
ζ is the damping ratio.

This equation is shown graphically in Figure 7.6.

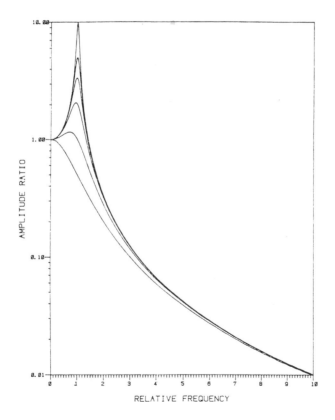

Figure 7.6   *Dynamic magnification factor*

The force transmissibility, T, of the system is defined as:

$$T = \frac{\text{force transmitted by the system}}{\text{force applied to the system}}$$

and is given by:

$$T = \left[\frac{[1+(2\zeta\omega/\omega_n)^2]}{[(1-(\omega/\omega_n)^2)^2+(2\zeta\omega/\omega_n)^2]} \right]^{0.5} \tag{7.19}$$

where the symbols are as for equation 7.18. This equation is shown graphically in Figure 7.7. For a system with negligible damping this reduces to:

$$T = 1/[(\omega/\omega_n)^2 - 1]^{0.5} \tag{7.20}$$

A value of $T = 1$ represents all the applied force being transmitted, and if $\omega = \omega_n$ the system is in resonance and the transmissibility tends to infinity.

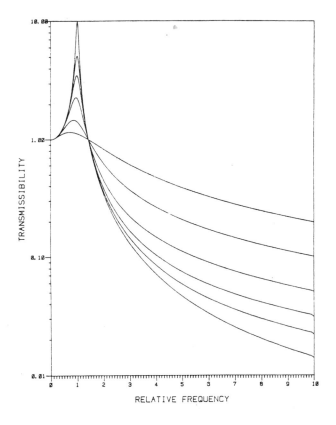

Figure 7.7   *Transmissibility*

The isolation efficiency of a mounting is therefore:

$$\% \text{ Isolation} = (1 - T) \times 100\%$$

Design of machine mountings

Machine mountings should be designed to reduce the force transmitted to the structure to acceptable levels for the particular application. Such mountings generally consist of a rigid mounting frame supported from the structure by resilient components such as steel springs or slab materials. Design of the mounting will generally depend upon how important it is that noise and vibration should not be transmitted to the surrounding areas. Critical areas are those where noise intrusion would be a severe problem, such as conference rooms or lecture theatres. While it is obviously not recommended to site noisy machinery close to these areas, it may be unavoidable in some situations. Where machines are mounted in critical areas, it may be necessary to achieve values of up to 97 per cent isolation or better, that is, a transmissibility of at least 0.03. Non-critical areas are those such as other plant rooms or machine tool workshops where intrusion in these areas may be as low as 70 per cent, that is, a transmissibility of 0.3.

Study of Figure 7.7 will reveal that significant isolation is not achieved until the ratio of the forcing frequency of the machine to the natural frequency of the mounting reaches a value of 2.5 to 3, representing isolation efficiencies of 80 to 90 per cent at low damping ratios. A good guideline is to aim for relative frequencies of 3 and above, although it should be noted that above this value little extra isolation is gained with increase in relative frequency.

It has already been mentioned that machines such as fans with separate motors should be mounted on the same baseplate or frame, and the whole assembly placed on resilient mountings. Since there may be disparity between the weights of the fan and motor, it will be necessary to determine the position of the centre of gravity of the assembly, which should be as near the centre of the baseplate as possible to ensure even distribution of the load on the mountings. This may be achieved by taking moments about the frame to determine the most suitable positions for the fan and motor. Should this prove difficult, taking moments about each of the selected mounting points will enable mounts of differing stiffnesses to be selected to distribute the load evenly. With complicated assemblies and/or in critical areas the use of a cast concrete inertia block may be advisable to help centralize the centre of the gravity. To achieve this such blocks should be about three times the weight of the machine. Machine and block are then considered to be one assembly, and resilient mounts selected as described below.

First, decide upon the required value of transmissibility, depending on whether the machine is to be mounted in a critical area. The forcing frequency of the machine will be known, usually the lowest rotational frequency of any moving parts, for a fan being the impeller frequency in revolutions per second (Hz). Depending on the type of isolator to be used, determine the damping ratio. This is of the order of 0–0.05 for steel springs, and 0.1 for rubber mounts. From Figure 7.7 determine the value of relative frequency necessary to achieve the required transmissibility at the chosen damping ratio, and hence determine the natural frequency of vibration of the mounting. Substitution of this value into equation 7.13 will give the total stiffness required, and equation 7.15 will yield the static deflection of each isolator. Selection of anti-vibration mountings may now be made from manufacturers' literature, and most manufacturers are willing to assist the engineer in selection if required.

Example A machine of mass 150kg is placed on a mounting of stiffness 1.65×10^6 N/m and damping 3.14 kNs/m and exerts a force of $625\sin 60\pi t$ N upon it. Calculate the damping ratio, the displacement amplitude, the transmissibility and the force transmitted to the structure.

$$\text{Forcing frequency, } \omega = 60\pi \text{ radians/second}$$

$$\text{Natural frequency, } \omega_n = [1.65 \times 10^6/150]^{0.5} = 104.88 \text{ rad/s}$$

$$\text{Relative frequency} = 60\pi/104.88 = 1.8$$

Damping ratio

$$\zeta = c/c_c$$

$$c_c = 2(km = 2)^{0.5}(1.65 \times 10^6 \times 150)^{0.5} = 31464 \text{ Ns/m}$$

$$\zeta = 3.14 \times 10^3/31464.3 = 0.1$$

Displacement amplitude

$$\text{Amplitude ratio} = x/(F_0/k)$$

$$F_0/k = 625/(1.65 \times 10^6) = 3.79 \times 10^{-4}$$

From equation 7.18, for relative frequency of 1.8 and damping ratio of 0.1, the amplitude ratio is calculated to be 0.44.

$$\text{Displacement amplitude} = 0.44 \times 3.79 \times 10^{-4} = 0.167 \text{mm}$$

Transmissibility

From equation 7.19, for relative frequency of 1.8 and damping ratio of 0.1, the transmissibility is calculated to be 0.47, which is high and would not be acceptable in most cases.

Transmitted force

> Transmissibility = Transmitted force/Applied force
>
> Maximum applied force = 625 N
>
> Transmitted force = $0.47 \times 625 = 293N$

Commercial anti-vibration mountings

A wide range of commercially produced mountings is available to suit most applications, with various static deflections and load-bearing capacities. The main types available are steel spring, rubber in shear and compression, rubber in shear or compression depending on the plane of mounting and various slab materials. Examples of such mountings are illustrated in Figures 7.8 and 7.9. Typical load ranges and static deflections are summarized in Table 7.9.

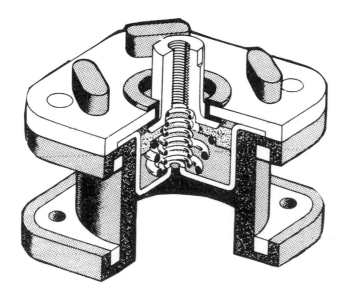

Figure 7.8   *Shock and vibration isolator*
(Reproduced by permission of Barry Controls.)

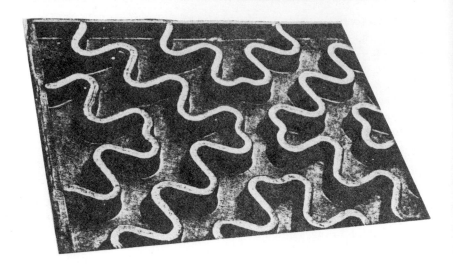

Figure 7.9   *Ribbed mat isolator*
(Reproduced by permission of Barry Controls.)

Table 7.9
Properties of commercial isolators

Type of mounting	Maximum load, kg	Static deflection, mm
Steel spring	2 000	50
	4 000	25
Rubber in shear and compression	1 600	6.5
Rubber in shear or compression	200	8.0
Slab compounds, studded	5 000	3.2/layer
ribbed	20 000	3.2/layer
compression stud	50 000	3.2/layer

References

1   Chartered Institution of Building Services Engineers (1973) *CIBSE Design Guide*, Section A1: Environmental Criteria for Design; Section B12: Sound Control.
2   American Society of Heating, Refrigeration and Airconditioning Engineers (1984) *ASHRAE Handbook: Systems Volume* chap. 32, Sound and Vibration Control.
3   Curle, N. (1955) *The Influence of Solid Boundaries upon Aerodynamic Sound*, Department of Mathematics, University of Manchester.

4   BS 848 (1985) Part 2: *Fan Noise Testing*, British Standards Institute, London.
5   Ver, I.L. (1984) Prediction of Sound Transmission through Duct Walls: Breakout and Pickup, *Ashrae Transactions*, Part 2A.
6   Brown, D.M. and Stewart, L.J. (1967) *Division of Noise in Branching Duct Systems*, HVRA Report No. 47, Heating and Ventilating Research Association.
7   Fairhall, D.M. (1987) Sound Reflection at Duct Terminations, *Proc. IOA*, Vol. 9, Part 1, February.
8   Fairhall, D.M., Roberts, J.P. and Vuillermoz, M.L. (1987) Attenuation at Side Openings in Air Ducts, *Proc. IOA*, Vol. 9, Part 2, April.
9   Heating and Ventilating Contractors' Association (1982) DW/142: *Specification for Sheet Metal Ductwork. Low, Medium and High Pressure/ Velocity Air Systems*, HVCA.
10 Holmes, M.J. (1973) *Air Flow Generated Noise*, Part 1: Grilles and Dampers, HVRA Report No. 75, Heating and Ventilation Research Association.

Chapter 8

APPLICATIONS FOR ACTIVE ATTENUATION

H.G. Leventhall, Head of the Institute of Environmental Engineering, and **John Roberts**, Acoustics Group, South Bank Polytechnic

Active cancellation of noise is the first genuinely innovative acoustic technique for more than a generation. Its emergence today as a practical possibility is due to the rapid progress made in computer hardware and the speed of digital signal processing. This chapter, then, has a somewhat different character from the others in the book. It describes the present state of the art of the application of active noise control, those areas where it may be successfully applied, the degree of likely attenuation to be attained and foreseeable future developments.

The essential principle of active attenuation (frequently called *anti-noise*) is shown in Figure 8.1, which illustrates the combination of two acoustic waves, an original and a second cancelling wave of the same frequency and approximately the same amplitude as the first but with nearly a 180° phase shift. For simple sine waves errors of about 10 per cent in the amplitudes and 10° in the phase of the second cancelling signal can still produce attenuations of 10 dB.

Where complex waves are to be cancelled the generated signal must be the exact opposite of the time waveform of the first, that is, contain the frequency components of the original waveform with the correct amplitude and phase relation. As will be shown, attenuation is most effective at lower frequencies and falls as the frequency increases.

For random noise, to give cancellation the signal must be closely (inversely) correlated with the original. A coherence of better than 0.95 is required to give worthwhile attenuation.

Active attenuation is usually considered only where conventional methods are too costly or bulky. Examples of possible applications are the cancellation

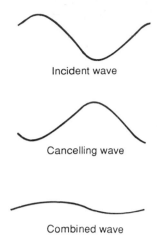

Incident wave

Cancelling wave

Combined wave

Figure 8.1   *Addition of waves showing incomplete cancellation*

of low frequency noise in small enclosures and ear defenders, in the ductwork of air-conditioning systems, and exterior noise radiated from engines, generators and transformers.

The theory and applications of active attenuation

The one-dimensional system

The one-dimensional system consists of a duct in which a plane wave (wavelength at least twice the larger cross-sectional duct dimension) is propagating. Active control applied in this situation has received most attention as it is the simplest application and has potential for large-scale use in air-conditioning systems.

Figure 8.2 shows schematic and corresponding block diagrams of the system. The primary noise – in most duct systems this will usually be generated by a fan – is detected by a microphone placed upstream of a secondary source (loudspeaker). The detected noise is then delayed to allow for the time the sound takes to travel the distance from microphone to loudspeaker. The electronic system must contain an amplifier for the microphone output and a suitable device, digital or analogue, to compensate for the acoustic characteristics of the duct and loudspeaker, to allow for the travel time of the sound down the duct from detecting microphone to loudspeaker and give the total phase shift of 180°.

With digital systems the minimum distance of the detecting microphone from the loudspeaker is determined by the speed of the microprocessor and

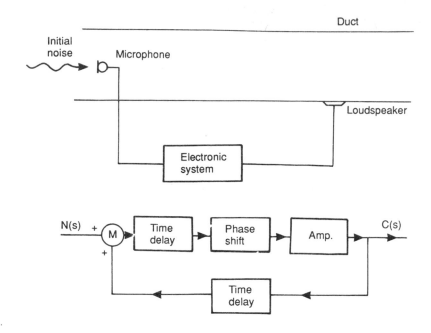

Figure 8.2   *Simplified schematic and block diagram of monopole active attenuator*

could be one or two metres. The loudspeakers will be mounted externally over holes cut in the walls of the duct, directly in the case of rectangular ducts but with the aid of a flange for circular ducts.

The frequency response of the loudspeaker in the duct is the least predictable of the system parameters and can be quite irregular. Experience has shown that the flexibility of digital control devices makes them easier to use than analogue systems, although the initial development may be more difficult.

The anti-noise signal radiates both up and downstream, producing a standing wave upstream and noise reduction downstream. The standing wave causes difficulties in the operation of the attenuator as it acts upon the microphone in a manner which is frequency dependent and tends to limit maximum attenuation to only a relatively small range of frequencies in the absence of suitable compensation.

The anti-noise system which is simplest to produce is the close-coupled, or tight-coupled attenuator developed by Hong, Eghtesadi and Leventhall,[1] shown in Figure 8.3. Here the microphone detecting the signal to be cancelled is situated in the duct sufficiently close to the loudspeaker generating the anti-noise for the time taken for the sound to travel from the speaker to microphone to be negligible compared with the period of the wave being cancelled.

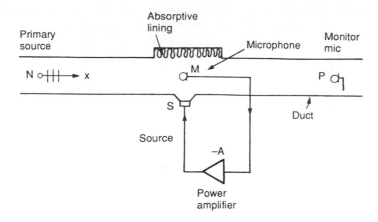

Figure 8.3 *Tight-coupled monopole in duct*

The system described above has achieved attenuations of up to 15 dB in the frequency range 200 to 300 Hz for a rectangular cross-sectional duct 450mm wide by 600mm high, as shown in Figure 8.4. Additional attenuation may be achieved by cascading two such devices.

Analysis of the tight-coupled system reveals that the non-linearities of loud-speaker response and the effect of the presence of the duct may be substantially compensated by setting the amplification as high as possible without provoking a feedback instability. This is achieved by gradually increasing the gain until a feedback whistle is obtained and then setting the gain back slightly.

The use of complex arrangements of microphones and sources to give absorption rather than redistribution of acoustic energy is also possible but the present trend is to use monopole systems and compensate for their deficiencies digitally. Self-adapting and self-monitoring systems capable of being set in

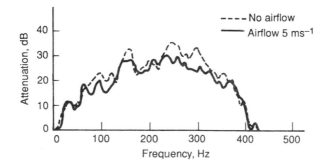

Figure 8.4 *Attenuation obtained using two cascaded tight-coupled monopole attenuators with high quality equipment*

the laboratory and adapting automatically on site are now being investigated and initial experiments have achieved 20 dB attenuation from 200 Hz to 500 Hz in an air-conditioning duct with a flow velocity of 10m/s.[2]

Fluid flows can give rise to turbulence and pseudo-sound which will detract from the performance of the active control system. Thus it is always advisable to avoid any fittings near the microphones which may generate turbulence, and in particular to place the initial detecting microphone sufficiently far from the fan (at least 2m) to avoid fan-generated flow turbulence.

Three-dimensional systems

Attenuation at the point of origin

If the noise source is localized, its radiation into open space can be reduced at low frequencies by converting the original monopole source into a dipole. This requires the installation of an opposite source close to the original. Acoustic dipoles are less efficient radiators of sound than monopoles when the spacing between the two sources comprising the dipole is small compared to the wavelength; there is thus a reduction in radiated sound energy at low frequencies. The acoustic field around a dipole is the well-known figure of eight pattern with cancellation along the normal bisector of the line joining the two sources.

The introduction of additional sources to transform the system into a quadrupole radiator will further reduce the total sound radiation, but the increased complexity makes such a system difficult to implement.

The simple monopole-to-dipole conversion has been applied, amongst other problems, to the open-end (300mm × 450mm) of a ventilation duct. A microphone is placed in the duct opening followed as closely as possible by a cancelling loudspeaker. Attenuation of 5 dB to 15 dB was achieved over the range 35 Hz to 250 Hz. Provided the end of the duct and the loudspeaker are sufficiently close the system has good potential.[3]

The dipole principle, of introducing a single negative source adjacent to the original, may be applied to practical noise problems if:

1   the original noise source is localized; and
2   the second source is placed much closer than one wavelength of the sound of interest to the original source.

While condition 1 requires a localized acoustic source it may be physically large, and the method has been successfully applied both to vehicle exhausts a few centimetres in diameter and gas turbine stacks a few metres in diameter. It is immediately obvious from condition 2 that this method will be particularly effective for low frequency noise. These conditions can also be met, for example, with noise from small machines such as small compressors.[3]

Active attenuation has been applied to control of noise from transformers, but large electrical transformer installations cannot be considered localized sources and noise reduction will be required all around the transformer rather than over a narrow area. The problem thus becomes the more complex one of control of sound from a radiating plane. The source to be controlled has a number of secondary sources placed around it (the more secondary sources the greater the possible sound reduction) and their outputs sequentially varied in amplitude and phase to reduce the total noise output to a minimum. Reductions of over 20 dB in the 100 Hz component and up to 12 dB in the overall level have been achieved by this method, but some care must be taken since with a limited number of secondary sources the original source cannot be perfectly matched and in some directions an actual increase in noise level may be experienced. As a general rule, to obtain more than 10 dB sound reduction from a large source secondary loudspeaker sources will need to be placed about half a wavelength apart.

Control of machinery noise is a natural extension of the process described above but with the further complication that the noise to be controlled may be relatively broad band. At present successful applications are limited to machines which produce regular, repetitive sounds.

Vehicle exhausts with their relatively small open area, low frequency noise and repetitive nature lend themselves readily to active attenuation and excellent results have been obtained in laboratory tests using the monopole-to-dipole conversion. Corr[4] employed an electronic pickup on the crankshaft to measure the repetition rate, digital processing to obtain the correct signal waveform, and division of the exhaust pipe into two, with two loudspeakers to give a quadrupole arrangement. Drive-by tests showed that the system achieved a reduction of some 7 dB at the troublesome lower frequencies. Before active control can be widely applied to vehicle exhaust noise it will be necessary to solve the problems of the use of microphones and loudspeakers in the aggressive environments associated with the commercial use of HGVs.

Some exhausts, such as those for the larger machines, will be of considerable size, for example several metres in diameter, and produce a high sound power output with wide-band and randomly varying acoustic outputs. The fundamental principle for attenuating the sound output of a large stack is exactly as for all other anti-noise devices: a wave of opposite nature is produced and mixed with the original noise. The differences arise from the physical size of the installation and the required power of the secondary wave. Sacks[5] describes a system using seventy-two loudspeakers with a total power input of 11kW located around the top of a turbine exhaust stack to give an attenuation of 10 dB from 22.5 Hz to 45 Hz with a maximum at 25 Hz.

Attenuation in enclosed spaces

Hearing protectors Under conditions of moderate noise exposure the requirements for hearing protectors can be easily met. But for conditions of intense low frequency sound or prolonged exposure the traditional passive earmuff has two serious disadvantages.

1 Low frequency performance is poor; even quality products may give only a few dB attenuation.
2 The seal around the ear is important and the clamping pressure of the headband can cause discomfort with prolonged wear, but a more comfortable, looser or more compliant seal will tend to detract from performance.

The solution to both these problems may be provided by active noise control.

Active attenuator ear defenders for personnel protection have been in use for some time and the basic technologies are now well understood. The attenuating system is normally some variation on the close-coupled monopole system (the size and geometry of the defenders do not lend themselves to other systems). Active systems can now be readily installed in conventional ear defenders to overcome poor low frequency performance. The Bose Corporation in the USA produces a headset for use in communicating with aircrew, which has given noise attenuation of 16 dB(A) over the range of 30 Hz to 1 kHz. These developments are now overcoming the need for a tight seal around the defender. A lightweight, comfortable open-back earphone headset has been developed and on industrial tests a cancellation of at least 10 dB was obtained over the frequency range 40 Hz to 600 Hz, with a maximum of 20 dB at 250 Hz.[6]

Noise-cancelling earphones for improved communication are, of course, very similar to ear defenders with active attenuation. Trinder and Jones[6] have developed the cancelling technique to reduce extraneous noise at the earpiece of a telephone. By injecting the electrical telephone signal into the cancelling system the telephone speech component can be accurately reproduced at the ear while, at the same time, the airborne noise component is cancelled. The result is a marked improvement in speech intelligibility.

There is clearly a place for noise-cancelling headsets in worker protection. The disadvantage of requiring the workers to be wired to external equipment or to carry all electronic and power supplies with them must be balanced against the improvement in performance of the defenders and the greater comfort during wear.

Enclosed spaces   Successful attenuation of noise in an enclosed space where the source is external has usually been confined to frequencies below the lowest modal frequency of the enclosure such that the sound distribution is fairly uniform throughout the space except close to the surface. The systems investigated have generally been experimental and are not yet ready for commercial application.

An early attempt to use active control to reduce noise levels in an enclosure was by Vian.[7] A directional microphone detected the noise entering an open window and, after suitable analogue processing, the signal was fed to a bank of nine loudspeakers distributed over the window area. At the centre of the room attenuation of up to 15 dB was achieved for low frequency pure tones and typically 5 dB in the range 200 Hz to 450 Hz.

Several workers have used the tight coupled monopole to provide attenuation within a small enclosure. This has been successful in producing substantial noise reduction in the region of the control microphone (up to 30 dB at 100 Hz) but the attenuation tends to be spatially restricted, falling off rapidly with distance. The accuracy of the signal processing controls the attenuation obtained. The next generation of rapid self-adapting systems should greatly enhance the attenuation.

Within an enclosed space the sound output of localized sources of noise has been reduced by monopole-to-dipole conversion. In one laboratory study[8] the original noise source was situated in one wall and a second source introduced immediately adjacent to it to provide a cancelling signal. A peak attenuation of 75 dB was obtained at 500 Hz, 500mm from the sources at the position of the detecting microphone. Attenuation of broad-band noise gave about 15 dB over the range 50 Hz to 400 Hz. The total attenuation to be achieved by this method is a function of the relative directivity patterns of original and secondary sources since these control the mixing of the original and cancelling waves.

An alternative approach is to establish secondary sources around a localized primary source within a space, in a similar manner to control of transformer noise. Experiments in an anechoic chamber and reverberant room using four secondary tripole sources achieved attenuation throughout the entire space exterior to the boundary defined by the secondary sources.[9] This latter method gives greater flexibility in the positioning of the secondary sources but they must still be close to the original noise source.

Physical separation of the noise and cancelling sources may be practical provided that they each strongly interact with the same principal mode of the enclosing space. The sound level generated by a primary source placed in the corner of a small enclosure to excite the (1,1,0) mode of 680 Hz was reduced by 16 dB when the cancelling source was placed immediately adjacent to it. When the secondary source was moved to the opposite corner – a position that gave good coupling with the excited mode – attenuation of 11 dB was obtained. In this position the secondary source changes the impedance into which the original source feeds but might also be considered to be 'reflected' acoustically into the corner with the primary source.

This method has been developed further by Elliott and Nelson[10] and applied to the interior of the cab of a sports car. Using adaptive filtering with two loudspeakers as secondary sources driven so as to minimize the sum of the squares of the acoustic pressure at four microphones situated near the driver's head, reductions of some 30 dB were achieved in the troublesome 100 Hz to 200 Hz engine firing frequency range.

Limitations of active control systems

The limitations imposed on the spatial reproduction of complex sound fields for active attenuation are of three types.

1 *Spatial limitations.* The cancelling source must be close to the original noise source. Active control of sound within enclosures is, as yet, truly effective only where the largest dimension of the enclosed volume is much less than one wavelength of the sound to be cancelled. Attenuation may be achieved over small regions in larger enclosures but generally only with costly three-dimensional systems.
2 *Frequency limitations.* Dipole and multipole cancellation is possible only over the frequency range for which the wavelength is long compared with the separation between the primary and secondary sources. In ducts there is also a strong frequency limitation imposed by the duct dimensions which determine the cut-off for plane wave propagation. Faster and more complex control procedures will be required if the frequency range is to be extended above 500 Hz, which often now forms the upper bound.
3 *Transducers.* Anti-noise systems tend to use readily available, general-purpose loudspeakers designed for use under relatively favourable conditions. Such transducers lack the robustness necessary for the relatively aggressive environments found in air-conditioning ducts or exhaust systems.

Loudspeakers for active attenuation

The cancelling sound source for active attenuation is normally a standard loudspeaker which must be capable of producing the same level as that from the noise source. An indication of the ability of loudspeakers to cope with the demands which active attenuation places upon them is given in the following hypothetical cases.

Suppose that the sound intensity level in a one-metre-square duct is 100 dB. This corresponds with a sound power in the duct of 0.01W, that is, an intensity of 0.01W/m². A loudspeaker operating at, say, 1 per cent efficiency requires only 1W of power amplification, which is easily obtained. A gas turbine exhaust may produce an acoustic power of 130 dB at the outlet of a passive silencer, concentrated in the low frequencies and corresponding to 10W of acoustic power. At 1 per cent efficiency this requires 1kW of power amplification. However, measures such as 100 dB or 130 dB are normally *averaged* levels, for example, rms, and the peaks in random noise may be 10 dB higher than this, requiring ten times more power amplification to control the peak noise.

Consider a typical bass loudspeaker with effective area A and an rms displacement of d. It can be shown that the power output level at frequency f is given by

$$L_w = 40\lg f + 20\lg(Ad) + 117 \qquad \text{dB re 1 pW}$$

Taking A as typically 0.07m² (300mm effective diameter) and d as typically 0.001m (1mm rms displacement) the sound power output at 63 Hz is

$$L_W = 106 \text{ dB } 106 \qquad \text{dB re 1 pW}$$

This shows that a single loudspeaker could be used as the cancelling source in the duct described above. However, it also follows that to achieve 130 dB sound power output at the 63 Hz frequency, about sixteen loudspeakers will be required. To allow for the peaks of 10 dB above the rms level the number of speakers must be increased to about fifty. It is seen that active attenuation for large sources is possible but requires a large amount of hardware.

The prospects for active attenuation

In 1988 active attenuation systems are available for the following applications:

1   noise to the atmosphere, for example exhaust noise;
2   ear defenders;
3   engine silencers; and
4   duct silencers.

Applications of active control systems to reduce noise radiated to the atmosphere have been specially tailored to fit the particular problem and each application is still of a 'one-off' variety. The other applications have the potential for wide-scale use and correspondingly large volume production. Once designers have been convinced of their viability they could become accepted elements of noise control systems.

A particular advantage of applications 3 and 4 is their potential for energy conservation. An active attenuator system may be used to provide the low frequency attenuation in an air duct without the costly pressure drop a traditional passive silencer would introduce. The savings in power consumption and increase in energy efficiency can more than balance the initial differences in capital and installation costs.[11] One American company retails a simple active attenuator control system for as little as £4000, but prices can be expected to fall with increasing production runs.

As the most significant development for many years, active attenuation has a clear place in noise control. Its applications are still restricted but there is continuing development and a prospective user should obtain realistic and informed advice before embarking on an active attenuation project.

Where can active attenuation be used?

Pure tones

We start with consideration of a pure tone because in such cases the positioning of the sensing microphones and loudspeaker(s) requires special consideration.

The noise radiated from the open end of a duct terminating in a factory wall is causing complaints from local residents. The noise is predominantly a pure tone of frequency 120 Hz, and the duct is rectangular, 600mm x 500mm carrying exhaust air at room temperature. The intensity of the tone measured in the duct at the proposed position of the loudspeaker is 120 dB re 10^{-12} W/m^2. Is the problem suitable for the application of active noise control techniques?

1   What *attenuation* is required? Active systems should give at least 10 dB and possibly as much as 20 dB for pure tones. Is such attenuation sufficient?
2   The *frequency* of concern is 120 Hz, well within the range over which active attenuation is usually effective. The 120 Hz signal has a wavelength of about 2.9m which is more than double the larger dimension of the duct and so is well within the frequency limits imposed by the duct size.
3   The *loudspeaker* power required is likely to be about 30W (sound intensity 1W/m^2, duct area 0.3m^2 and loudspeaker efficiency about 1 per cent). This is readily achievable.

This problem thus appears well suited for the application of active control.

The pure tone is likely to give rise to a standing wave in the duct and some care must be taken in the positioning of the loudspeaker(s) and microphones. The loudspeaker(s) should be placed at a pressure maximum (antinode) to ensure maximum impedance matching and the best conditions for sound output from the speaker. The open end of the duct will be a pressure minimum (node) and, allowing for end corrections, the loudspeaker(s) should be placed at a distance of 1.9m, 3.4m, or 4.8m from the duct termination.

The initial and compensating microphones will be placed upstream and downstream respectively of the loudspeaker. The travelling wave will be partly reflected at the open end of the duct to produce a standing wave. It is usual to place the initial microphone at a pressure node (minimum) of this standing wave to obtain maximum sensitivity to the travelling wave (which after all is what is to be cancelled), and to place the error or compensating microphone at a pressure antinode to maximize the response to any standing wave – that is, to maximize the response to the remnant of the tone getting past the loudspeaker(s).

If the loudspeaker(s) were placed 3.4m from the open end of duct, the initial microphone would have to be placed at least 2m to 3m upstream. It is thus convenient to place the initial microphone at one of the pressure antinodes occurring at either 5.55m or 7.0m from the open end of the duct.

The downstream, error, microphone would then be placed at the pressure antinode 1.9m from the open end of the duct.

Wide-band noise

A combustion system produces broad-band combustion roar centred on the 250 Hz octave band. It is predicted that the use of standard passive noise reduction techniques will preferentially reduce the higher frequencies leaving a broad noise spectrum centred on the 63 Hz octave band. Is this residual noise problem amenable to active attenuation?

1   What *attenuation* is required? We have seen (for example Figure 8.4) that the likely broad-band attenuation will be:

Mid-frequency octave band, Hz	31	63	125	250	500
Likely attenuation, dB	8	10	15	10	0

Are these attenuations likely to be sufficient?
2   The *frequency* range appears suitable for treatment by active methods, but the upper limit must be found from the wavelength corresponding to twice the diameter of the flue.
3   The *loudspeaker* requirements must be determined from measurement of the noise travelling up the flue. Is the required loudspeaker power practicable?

If these questions are answered positively active attenuation might form part of the total noise reduction techniques used.

With wide-band noise and the absence of pure tones the positioning of the microphones and loudspeaker(s) is much less critical. The initial microphone must lead the loudspeaker(s) by at least 1m to 2m and the error microphone must follow the loudspeaker but it can be placed reasonably close, say 300mm.

References

1   Hong, W., Eghtesadi, Kh. and Leventhall, H. (1983) The Tight-Coupled Monopole Active Attenuator, *Noise Control Engineering*, Vol.22, No.1, pp. 16–20.
2   Roure, A. (1985) Self-Adaptive Broadband Sound Control System, *J. Sound and Vibration*, Vol.101, pp. 429–41.
3   Short, W, (1980) Global Low Frequency Active Attenuation, *Proc. Internoise '80*, pp. 694–8.
4   Corr, H. (1986) Active Control of Exhaust Noise from a Commercial Vehicle, *Proc. IOA*, Vol.8, Part 1, pp. 159–64.

5   Sacks, T. (1982) Killing Noise with Noise: A Sound Solution, *Electrical Review*, Vol.221, No.17. 19 November, pp. 22–3.

6   Trinder, M. and Jones, O. (1987) Active Noise Control at the Ear, *Noise-Con '87*, pp. 393–8.

7   Vian, J. (1977) Elimination du Bruit par Absorption Active, *Rev. d'Acoustique*, No.43, pp. 322–34.

8   Warnaka, G. *et al.* (1983) Active Control of Noise in Interior Spaces, *Proc. Internoise '83*, pp.415–18.

9   Sha Jia-zheng and Tian Jing (1986) Active Noise Attenuation in Three-Dimensional Space, *Proc. 12th ICA,* Paper C5–6.

10 Elliott, S. and Nelson, P. (1987) The Active Control of Enclosed Sound Fields, *Proc. Noise-Con '87*, pp.359–64.

11 Hong, W., Eghtesadi, Kh. and Leventhall, H. (1985) Energy Conservation by Active Noise Attenuation in Ducts, *Proc. IOA*, Vol.7, Part 2, pp.103–9.

Appendix

Vibrating systems

The vibrating system described in Chapter 1 may be analysed in the following way. The driving force $F = F_0 \cos \omega t$ is first written in complex notation

$$F = F_0 \exp(j\omega t) \tag{A.1}$$

where $j = \sqrt{-1}$.

In this notation the periodic force is represented as a vector of length F_0 which is made to rotate, like the spoke of a wheel, in an anticlockwise direction with an angular frequency ω radians per second. The sinusoidal function then becomes the projection of this vector along one of the axes as time t progresses – see Figure A.1. By convention the x and y axes are called the real and imaginary axes respectively, and as the full function contains components in both directions it represents rather more than the simple cosine function given above. For the following analysis it is sufficient to treat the expression as if it were the cosine function alone on the understanding that we take only the real part of any solution. The particular advantage of this notation is that it enables sinusoidal functions to be readily differentiated or integrated as follows:

$$\begin{aligned}
&F = F_0 \exp(j\omega t) \\
&dF/dt = j\omega F_0 \exp(j\omega t) = j\omega F \\
&\textstyle\int F dt = (1/j\omega) F_0 \exp(j\omega t) = F/j\omega
\end{aligned} \tag{A.2}$$

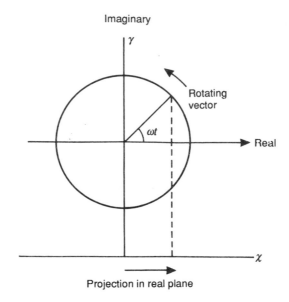

Figure A.1   *Sinusoidal oscillation in complex notation*

and similarly for sinusoidal functions of other quantities, such as displacement and pressure.

For the problem of the mass on a spring, we may obtain its response to the driving force by applying Newton's second law of motion. This tells us that net force is equal to mass times acceleration, so that taking into account the applied force and the two opposing forces we may write

$$F - s\xi - r(\mathrm{d}\xi/\mathrm{d}t) = m(\mathrm{d}^2\xi/\mathrm{d}t^2) \tag{A.3}$$

This equation may be rewritten in terms of velocity as

$$m(\mathrm{d}u/\mathrm{d}t) + ru + s\int u\,\mathrm{d}t = F \tag{A.4}$$

We now assume that the solution to this equation is a sinusoidal velocity of the same angular frequency as the force and of the form

$$u = u_0\exp[j(\omega t + \theta)] \tag{A.5}$$

θ is an arbitrary phase angle which must be included at this stage as velocity and force need not necessarily be in phase.

From this solution we obtain $\mathrm{d}u/\mathrm{d}t = j\omega u$, and $\int u\,\mathrm{d}t = (1/j\omega)u$ and on substitution equation A.4 becomes

$$j\omega m u + ru + su/j\omega = F \tag{A.6}$$

or

$$u = F/[r + j(\omega m - s/\omega)] \tag{A.7}$$

The quantity $r + j(\omega m - s/\omega)$, which determines the instantaneous velocity of the system, is called the *complex mechanical impedance*, Z_m, and the term in brackets the *reactance*. In the form given it contains information on both phase and amplitude but for most practical applications we would be concerned with the amplitude only and this may be obtained from the modulus of the impedance

$$|Z_m| = [r^2 + (\omega m - s/\omega)^2]^{0.5}$$

Hence

$$u = \frac{F}{[r^2 + (\omega m - s/\omega)^2]^{0.5}} = \frac{F}{|Z_m|} \tag{A.8}$$

where u and F would represent simultaneously the peak or root mean square values of the relevant sinusoidal functions.

When dealing with the behavior of systems at or near resonance it is frequently helpful to be able to describe the way the velocity amplitude changes as the frequency moves across the resonant peak. This, the shape of the resonant curve, may be quantified by what is called the Q factor of the system. Q can be defined as the ratio of the resonant frequency to the bandwidth, where the bandwidth is the difference between the two frequencies at which the velocity falls to $1/\sqrt{2}$ of its value at resonance on either side of the peak. The points are shown in Figure 1.2 and are referred to as the half-power or 3 dB points. We therefore have

$$Q = \omega_0/(\omega_2 - \omega_1) \tag{A.9}$$

The sharper the resonant peak, the higher will be the value of Q.

It may be shown using equation A.7 that provided the resistance is less than, say, one-tenth of the reactive component

$$Q = \omega_0 m/r \tag{A.10}$$

This relationship is useful as it enables some measure of the resistance of a system to be obtained directly from the resonant curve without the need to know the magnitude of the driving force.

The wave equation

The equation for the plane travelling wave was given in Chapter 1 as

$$\xi = \xi_0 \exp[j(\omega t - kx)] \qquad\qquad (A.11)$$

Instead of deriving this formula from first principles we will attempt to establish it by looking at its physical interpretation. To do this it is first necessary to think of all directions in space as consisting of two components, one which is real and which we can have knowledge of, and the other which is imaginary and which could never be observed. It is as if we lived in a two-dimensional world and could see only shadows of objects and not components of their length along the direction of the light beam casting the shadow. It follows that the imaginary direction would be at right angles to the real in this complex space.

In such a space the factor $\exp(j\theta)$ is a mathematical operator or instruction to rotate a vector through an angle θ in an anticlockwise direction. The formula for the wave equation A.11, which may be written

$$\xi = \xi_0 \exp(j\omega t)\exp(-jkx) \qquad\qquad (A.12)$$

contains two such instructions which we are able to consider separately. The vector ξ (called the *displacement amplitude of the wave*) will at the origin $(x=0)$ and at the start of time $(t=0)$ stand in the real plane. If we now consider movement along the x axis, that is, along the direction of travel of the wave, the operator $\exp(-jkx)$ rotates the vector into imaginary space (clockwise because of the minus sign) through an angle kx. After a distance such that $kx = 2\pi$, the vector will have rotated through a complete circle and be back in the real plane.

$\xi_0\exp(-jkx)$ therefore describes a spiral in complex space which repeats itself over a distance $2\pi/k$, a distance which is of course, equal to the wavelength, as shown in Figure A.2. The second operator $\exp(j\omega t)$ instructs

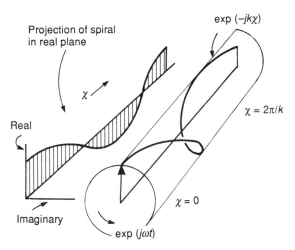

Figure A.2   *Complex representation of a travelling wave*

the whole spiral to rotate anticlockwise as time progresses with an angular frequency ω. Finally, because we can see only the real component of the complex vector, the original travelling wave will be the shadow or projection of this spiral in the real plane as it rotates.

The velocity of sound

The equation for the plane travelling wave may now be used to find the velocity of sound. This is done by differentiating the plane wave equation twice with respect to time and then twice with respect to distance, using the method explained previously, to obtain the differential wave equation as follows:

$$\partial^2 \xi / \partial t^2 = -\omega^2 \xi$$

and

$$\partial^2 \xi / \partial x^2 = -k^2 \xi$$

(A.13)

so that on equating the two displacements and with some rearrangement we obtain

$$\partial^2 \xi / \partial t^2 = (\omega/k)^2 (\partial^2 \xi / \partial x^2)$$

(A.14)

But c is equal to ω/k, so the equation may be rewritten

$$\partial^2 \xi / \partial t^2 = c^2 (\partial^2 \xi / \partial x^2)$$

(A.15)

This is the general form of the one-dimensional differential wave equation of which the equation for the plane travelling wave is just one solution.

The importance of the differential equation is that it shows the essential physical properties of the medium through which the wave may travel, and from it we may obtain the wave velocity as shown below.

Consider air at atmospheric pressure contained in a uniform pipe Figure A.3. If a small element of this air of length δx is displaced from its rest position so that one end moves through a distance ξ and the other end through a distance $\xi + \delta \xi$, there will be a change in the volume of the element of gas and hence a change of pressure. The magnitude of this pressure change, that is, the pressure above or below atmospheric, is given by the bulk modulus of the gas, K, multiplied by the fractional change of volume, so that

$$p = -K(\partial \xi / \partial x)$$

(A.16)

This means that there will be a pressure variation in the medium only if

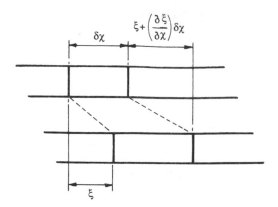

Figure A.3   *Displacement of air in a sound wave*

there is a variation in the rate of change of displacement with distance.

For sound waves we know that such variations do occur over the wavelength, so there will be pressure variations in the air as the sound wave passes through it and these in turn will create forces which act on the elements of the gas causing them to accelerate.

The magnitude of this acceleration may be obtained from Newton's law and the pressure difference across the element as follows. If p is the pressure at one end of the element then that at the other will be $p + (\partial p/\partial x)\delta x$, so that the difference $-(\partial p/\partial x)\delta x$ will be the force per unit cross-sectional area acting on the element. The acceleration is then given by

$$-(\partial p/\partial x)\,\delta x = -\rho\delta x(\partial^2 x/\partial t^2) \tag{A.17}$$

but

$$p = -K(\partial \xi/\partial x)$$

so that

$$\partial^2 \xi/\partial t^2 = (K/\rho)\,(\partial^2 \xi/\partial x^2) \tag{A.18}$$

Comparing this equation with equation A.15 we obtain the important result that the velocity of sound in air is given by

$$c = (K/\rho)^{0.5} \tag{A.19}$$

To complete this expression it is necessary to accept that the changes in volume and pressure take place under adiabatic conditions so that the bulk modulus K is γP. γ is the principal specific heat ratio (1.4 for air), and P is the atmospheric pressure. The wave velocity then becomes

$$c = (\gamma P/\rho)^{0.5} \qquad (A.20)$$

Sound intensity

Sound waves may be considered to transport energy in the following way.

Take a small area dS at right angles to the direction of propagation of the wave; this area may be regarded simply as a single layer of air molecules. At any instant a force will act on this layer equal to the product of the acoustic pressure and the surface area. If this layer remained stationary no work would be done but we know that the molecules are being continuously displaced back and forth about their equilibrium position so that the point of application of the force will move.

If such movement takes place over a distance $d\xi$ and in a time dt, the work done (force times distance) will be $p\, dS.\, d\xi$, and the power (rate of doing work) will be this work divided by the time, that is

$$W = p\, dS\, (d\xi/dt) \qquad (A.21)$$

But $d\xi/dt$ is the particle velocity, u, so that the power per unit area or intensity will be given by

$$I = W/dS = pu \qquad (A.22)$$

This equation shows that the instantaneous rate of transmission of energy from one layer of air to the next in the sound wave is the product of the acoustic pressure and the particle velocity.

Sound radiation

When a sound source radiates into free space the impedance into which it feeds varies with the ratio of the dimensions of the source and the sound wavelength.

A parameter which is frequently chosen to represent this relationship between the source dimension and the wavelength is kD, where $k = 2\pi/\lambda$ and D is the source diameter, and the radiation impedance may be expressed in terms of this variable as

$$Z_R = (\rho c/S)[R(kD) + jX(kD)] \qquad (A.23)$$

where $R(kD)$ is called the *piston resistance function* and $X(kD)$ the *piston reactance function*. The radiated power W_r would then be given by

$$W_r = (Su)^2\, (\rho c/S)R(kD) \qquad (A.24)$$

The full expressions for these functions are fairly complicated but for most practical applications it is sufficient to consider them for the two extreme conditions only, that is, when the wavelength is much larger than the source dimensions and when it is much smaller. For a piston operating in a baffle so that it radiates only into one hemisphere these functions are

$$R(kD) \approx 1/8k^2D^2 \quad \Big\} \qquad kD < 1 \qquad \text{(A.25)}$$
$$X(kD) \approx 4kD/3\pi$$
$$R(kD) \approx 1 \qquad \Big\} \qquad kD > 4 \qquad \text{(A.26)}$$
$$X(kD) \approx 0$$

This shows that for a small source emitting low frequency sound the radiated power will be proportional to k^2 which is the same as being proportional to ω^2, so that at low frequencies the power can be very small indeed. This is of course why low frequency loudspeakers need to be so large to provide a reasonable output.

This also shows that with a finite reactive component $X(kD)$ at low frequencies there will be a component of velocity 90° out of phase with the pressure. This is equivalent to a volume of air moving back and forth with the loudspeaker diaphragm which although not contributing to the radiated power may be shown to be responsible for the so-called 'end correction', usually associated with resonance tubes.

At the other extreme we see that the impedance is real, as $X(kD) = 0$, so that p and u are in phase and that its magnitude is the same as if the source were radiating into a uniform tube. This is because at high frequencies the radiated sound energy becomes more confined to the forward direction and approximates to a parallel beam as if in a tube.

It is of interest to note that the change of radiation impedance with frequency when kD is small is used by loudspeaker manufacturers to provide a region in which the loudspeaker response can be made very nearly uniform with frequency. If the speaker as a vibrating system is made to operate above its resonant frequency, there will be a range over which the fall in velocity with frequency just matches the corresponding increase in radiation resistance to give a uniform power output. Unfortunately for practical speakers the region in which this can occur is not very wide in the frequency scale and more than one speaker is usually needed to cover the full spread of the audio spectrum.

Although sources in general may differ in detail from the idealized case considered above they will all tend to display the same general trends when radiating into free space.

Index

*Note: Figures in italics refer to illustrations and tables

AAD 96–7
Aberdeen Airport 68
absorbers, suspended 164–5
accelerometer 34
  piezoelectric 37–8
acoustic absorption 161–5
  increasing 164–5
  prediction of performance 162–4
acoustic booths 152–3
acoustic intensity analysis 33–4
acoustic measurements 27–34
acoustic sprays 165
active attenuation 226–38
  in enclosed spaces 231–3
  limitations of control systems 234
  loudspeakers for 234–5
  one-dimensional system 227–30
  at point of origin 230–1
  prospects for 235
  theory and applications of 227–33
  three-dimensional systems 230–1
  use of 236–7
  and pure tones 236–7
  wide-band noise 237
aerodynamic noise 142–4
aggregate adverse deviation 95, 96–7
air capacitor microphone 34, 35–7
air distribution systems 192–214

airborne insulation ratings 95
airborne noise 90, 91
airborne rating system (STC) 96
airborne sound insulation indices 95
aircraft noise 48, 49, 64–8
  insulation 64–5
  planning 65
alarms, audible intruder 84
Allen formula 199
American Society of Heating,
      Refrigeration and Airconditioning
      Engineers
  (ASHRAE) 138, 192
  Handbook 201, 205
amplitude ratio 219
anechoic chamber 34
anti-noise see active attenuation
anti-vibration isolators 223, 224
anti-vibration mountings, commercial
      223, 224
antinodes 11
Approved Documents 133
assumed protection 177
attenuator, sound
  active monopole 228
  tight-coupled 228, 229
  two cascaded tight-coupled 229
audiometry

employees' access to results of 190
in hearing conservation programmes
183–5
interpretation of 187–9
tests
audiometric booth for *186*
conditions for 185–7
headphones for *186*
Audiometry in Industry (HSE) 188

basic noise level (BNL) 56
'boilermakers' deafness' 168
booster noise 144
British Airports Authority 64
British Gas 138
British Rail 68, 69, 72
British Standards
BS848 Part 2 195
BS4142 76–6, 78–80
BS5108 175
BS5228 74, 75
BS6472 63
Building Regulations (1985) 97, 133, 134
Building Research Station 76
Building Services Research and
Information Association 205
burner silencer *145*

Calculation of Road Traffic Noise
(CRTN) 56, 57, 61
calibrator, electroacoustic 29
factory testing of 32
capacitor microphone cartridge 36
cathode ray tube (CRT) 45
cavitation 215, 216
characteristic impedance 6–7
Chartered Institution of Building
Services Engineers (CIBSE) 138
Guide 203
Civil Aviation Act 64
Civil Aviation Authority 66, 67, 68
clay pigeon shooting 84
coincidence effect 18, *19*
combustion driven resonance 145
common law of nuisance 134
complex mechanical impedance *241*
construction noise 73–6
constructional techniques 108–21
principles 108–10
absorption 110
discontinuities 109
mass 108–9
resilience 109–10

stiffness 109
uniformity 109
types of construction 110–21
continuity of velocity 17
continuously welded rail (CWR) 69–70
Control of Pollution Act (1974) 74, 75,
80, 84, 134–5
corrected noise level (CNL), 77, 78
coupler 29
crest factor 39
critical damping 218
critical frequency 18
Curle's equation 204

daily acoustic immission 169
damped systems 218
damping ratio 222
decibel 19
decibel 19–21
Department of the Environment Circular
10/73 Planning and Noise 58, 65,
71
Department of Transport 64, 68
diesel multiple units (DMU) 69
digital tape recording 44
dipole 9, 231
direct transmission 90, 92
directivity 209
Disco rules OK? 83
discotheques 83–4
displacement amplitude 222, 242
dose meters 171
duct systems
attenuation in 201–4
at divisions of flow 201–3
at elbows and bends 201
in mitred bends *202*
in straight sheet metal 201, 202
crosstalk noise 200
duct break-in noise 198
duct break-out noise 198, 199–200
duct linings 212–13
end-reflection losses 203–4
expansion plenum chambers 213–14
noise control in 209–14
acoustic treatment 210–11
packaged silencers 211
system design and installation 209–
10
and room sound levels, estimation of
208–9
velocity-generated noise 204–8
dynamic magnification factor 219

earmuffs 179, 181
  headphone-style 174
earplugs 181
  custom-made individually moulded 174
  single-use, disposable standard-sized
    173
  special expanding foam plastic
    universal fit 174
eigentones 12
electret microphone 37
electric multiple units (EMU) 69
electroacoustic calibrator 29
enclosures 146–54
  acoustic booths 152–3
  alternative methods of 152–4
  complete 146–52
    access 148
    construction 146–8
    mechanical isolation 150–2
    prediction of performance 148–50
  influence of 10–11
  partial 153–4
  tumbler 151
  underground installation 154
entertainment noise 80–85
  discotheques 83–4
  finding the correct balance 85
  pop concerts 80–2
environmental noise analysers 42
Environmental Noise and Vibration,
  Guidelines for (GLC) 58–9
equal loudness contour 23, 24
equivalent continuous sound pressure
  level, L_{eq} 26, 68, 169, *170*, 171
European Council Directive 86/188/EEC
  137, 169
Evaluation of Human Exposure to
  Vibration in Buildings (BS6472/
  1984) 63

fan noise 144, 193–204
  and correction factors *196*
  and correction for fan efficiency *197*
  estimation procedures 195–7
  and installation 198
  manufacturers' data 195
  sound power output 195–7
  see also duct systems
Fan Noise Testing (BS 848 Part 2) 195
Fast Fourier Transform (FFT) analyser
  25
Federal Housing Administration 96
flanking transmission 90, 92–3, 200
flexural waves, 18, *19*

'footfall' 91
forcing frequency 216
Framing Noise Legislation (HSC) 169
free field measurements 33
frequency bands 24–6
frequency of forced vibration 216

Gatwick Airport 67, 68
Gauss's theorem 34
GLC 66, 71, 74, 80, 82
  Guidelines for Environmental Noise
    and Vibration 58–9
graphic level recorders 43

Health and Safety Commission (HSC)
    169, 188
  Audiometry in Industry 188
  Framing Noise Legislation 169
hearing conservation programmes 168–91
  audiometry 183–5
    interpretation of the audiogram 187–
      9
    test conditions for 185–7
  criteria 168–71
  educational needs 182–3
  surveys 171–3
hearing protection and protectors
  and active attenuation 231–2
  and adequate sound attenuation 175–
    81
  choice of 175–82
    assumed protection of 178
    and compatibility with other head-
      worn items 179
    and compatibility with other
      safety and hygiene equipment 179
  comfort and hygiene 180–1
  communication and general safety
    needs and 180
  cost of 181–2
  performance 175
  types of 173–4
hearing protectors *see* hearing protection
    and protectors
Heathrow Airport 48, 64, 67, 68
Heating and Ventilation Research
    Association 205
helicopter noise 48, 66
Helmholtz resonators 145
high pressure exhaust muffler 142, *143*

ice cream chimes 84
'impact' 91

impact insulation ratings 95–6
impact isolation class (IIC) 96
impact noise ratings (INR) 96, 97
impact sound 141–2
impact sound insulation index 95
impact sound pressure levels 94, 95–6
impact transmission levels 94
incomplete cancellation of waves *227*
industrial noise 76–9
  assessing likelihood of complaints 78
  background noise level 77–8
  and BS4142 76–7
    problems of BS4142 78–9
  from industrial combustion systems
    144–6
  measuring factory noise 77
  scope for improvement 79
infrasound 25
insertion loss 149
instrument selection 41–2
instrumentation 34–45
International Civil Aviation Organization
    (ICAO) 66
  Committee on Aircraft Noise (CAN)
    66–7
international rating system (ISO) 96
intruder alarms 84
isolators, anti-vibration 223, *224*

'knocking' noises 216

Land Compensation Act (1973) 57, 73–4
legal considerations of sound insulation
    in buildings
  common law of nuisance 134
  new-build 133
  prevention 132–3
  rehabilitation 133–4
  remedy 134–5
liquid crystal display (LCD) 45
Luton Airport 68

machine mountings, design of 221–2
machinery vibration and isolation 216–23
  damped systems 218
  and design of machine mountings
    221–2
  undamped systems 217–18
management
  work processes 166
  work procedures 166
  workshop practices 165–6
Manchester Airport 68
mass law 18
mass-controlled regions 4

maximum permissible exposure level
    (MPEL) 83
measurements, acoustic
  and factory calibration 32
  free field 33
  premeasurement checks 29–32
  and sound intensity 33–4
  sound power 33
  reverberant room 33
mechanical impedance 3
Method of Rating Industrial Noise
    Affecting Mixed Residential and
    Industrial Areas (BS4142) 76–7,
    78–80
model aircraft 84
motor sports 85
Motor Vehicles (Construction and Use)
    Regulations 59
moving coil meter 45

narrow band analysers 25
National Physical Laboratory 148
national frequency 217
new-build 130–1, 133
Newton's Second Law of Motion 240,
    244
nodes 11
Noise Advisory Council 70, 71
Noise and Number Index (NNI) 65, 66,
    67, 68
noise control
  by acoustic absorption 161–5
  building layout 138–9
  building materials 139
  at the design stage 138–40
  equipment and services 139–40
  by enclosures, partitions and screens
    146–61
  and good management 165–6
  an integrated approach 138
  reasons for 137–8
  reduction at source 141–6
  remedial 140–1
Noise Control on Construction and Open
    Sites (BS5228) 74, 75
noise criteria curves 192
noise dose 168, 171
noise dose meters 171
noise-induced hearing loss 47, *184*, 185
Noise Insulation Regulations (1975) 57,
    58, 61, 62, 73
noise protection helmet 174
noise rating 47–51
  individual response 50–1

quantifying 48
setting a standard 50
subjective reactive to 48–9
noise rating curves 192
Norris-Eyring formula 15
nuisance 134

octave band analysis 171
overdamping 216
partitions 154–60
composite 156–7
double-leaf 157–60
prediction of performance 155–6
percentile sound level, L_n 27
phon 24
piezoelectric accelerometer 37–8
piston reactance function 245
piston resistance function 245
pistonphone 29
plane waved 14
Planning and Noise (Circular 10/73), 58,
65, 71
plenum chambers 213–14
pop concerts 80–1
Pop Concerts, Code of Practice (GLC)
80–2
*Preevention of Damage to Hearing from
Noise at Work* 169
pump noise 145

railway noise 68–73
annoyance 72–3
criteria 71–2
locomotive noise 69
noise reduction 72
prediction 70–1
sources 68–9
wheel-rail noise 69–70
rating systems, sound insulation 94–6
airborne 95
equivalence between systems 96
impact 95–6
reactance 241
recorders 42–5
*Reduction of the Exposure of Employed
Persons to Noise, Code of Practice
for* 136, 169
rehabilitation 130–1, 133–4
remedial works 121–9, 131–2
assessing the problem 123
construction 124–5
design and layout 124
limitations 124–5

resonance 4
reverberant field 14
reverberant room measurements 33
reverberation 14–16
reverberation time 14
ribbed mat isolator *224*
road traffic noise 51–64, 73
character of 51–3
facade measurement of 54, 55
field measurement of 54, 55
legislation and standards 57–9
measurement of 53–6
prediction of 56–7
preventive measures 59–62
at the receiver 61–2
at source 59–60
between source and receiver 60–1
vibration 62–4
room sound levels 208–9
direct field 208–9
reverberant field 209
south fields 208–9
Royal Automobile Club 84
Sabine formula 15
Sabine theory 162
screens 160–1
secondary independent panels *116*, 127
sensitivity 36
shock isolator 223
silencers 143, *144, 145*, 211–12
single event exposure level 26
sinusoidal oscillation in complex notation
240
sound decay 14–16
sound exposure level (SEL), L_{AX} 26–7
sound insulation
design feedback 108
detail design 103–8
of floors 103–4, 110–15, 126–7
concrete 110–12, 126–7
platform 112, *113*, 126
raft 113, *114*
stairs 127
suspended 112–15
timber 126
insulation by design 99–108
of junctions 119–21
panel and duct 122
wall/floor *122*
layout design 100–3, 124
legal considerations 132–5
of openings 106–8
project definition 99–100
remedial works 121–9

of secondary independent ceiling 113–15
of service ducts 119, *120*
of service installations 105–6, 128–9
trouble-shooting 129–31
walls 104–5, 115–21, 127–8
  cavity 117–18
  compound 128
  masonry 127–8
  panel systems for 116
  solid 115–17
  timber-frame, 117, 128
  wall finishes, effect of 118–19
  see also sound insulation between
    dwellings
sound insulation between dwellings
    89–136
  assessing 93–8
  criteria and standards 93–4
  rating 94–6
  standards for 96–8
    guide values for 98, 99
    recommended by expert bodies 98
    required by client groups 98
  transmission paths 90–3
sound intensity 7–8, 23, 245
sound intensity analyser *34*
sound intensity measurements 33–4
sound level difference 95
sound level meter 23, 28, *30*, 38–41, 171
*Sound Levels in Discotheques, Code of
    Practice for* (HMSO) 83
sound pressure level 23, 27–9
sound power 23
  measurements 33–4
sound radiation 8–9, 245–6
  absorption in a room 13–14
  hemispherical 33
  and influence of the enclosure 10–11
  into a room 11–12
  spherical 33
sound reduction index 21, 95, 148, *200*
sound transmission class 95
sound transmission loss 94, 148
sound transmission through barriers 16–
    19
standing wave 11
Stansted Airport 67
Stanworth
  Railway Noise and the Environment 70
static deflection 217
stiffness-controlled regions 4
structural and non-structural flanking 90,
    92–3

structureborne noise 90, 91
suspended absorbers 164–5

tape recorders 43–5
'tapping machines' 91
temporary threshold shift (TTS) 185
tests, measurement
  'A' weighting 31
  between range linearity 31
  L_{AE} (SEL) 32
  L_{eq} or L_{Aeq} 31–2
  L_n and L_{An} 32
  time constants 31
  within range linearity 31
tight-coupled monopole attenuator 228,
    29
two cascaded *229*
Town and Country Planning Act (1971)
    66
'trade-off' formula 130
Traffic Noise Index 53
transducers 234
transmissibility 220, 222
transmission paths 90–3
transmitted force 223
trouble shooting 129–32
  common failures 129
  expected performance and costs 130–2
  preventing and remedying failures
    129–30
    during construction works 130
    post-occupation 130
    pre-construction works 129
    pre-occupation 130

ultrasound 25
underdamping 218

velocity of sound 243–5
vibrating systems 2–4, 239–41
vibration isolator 223
vibrations
  forced 218–19
  road-traffic induced 62–4

water distribution systems 214–16
  pipework noise 215
  pump noise 215
  valve noise 216
wave equation 4–6, 241–3
'weavers' deafness' 168
weightings 23–4
wide-band noise 145
Wilson Committee report 51, 67
working formulae 19–21